Audio for Television

For Roderick

Audio for Television

John Watkinson

Focal Press

OXFORD AMSTERDAM BOSTON LONDON NEW YORK PARIS
SAN DIEGO SAN FRANCISCO SINGAPORE SYDNEY TOKYO

Focal Press
An imprint of Elsevier Science
Linacre House, Jordan Hill, Oxford OX2 8DP
200 Wheeler Road, Burlington, MA 01803

First published 1997
Transferred to digital printing 2003

British Library Cataloguing in Publication Data
A catalogue record for this book is available from the British Library

Library of Congress Cataloguing in Publication Data
A catalogue record for this book is available from the Library of Congress

ISBN 0 240 51464 5

For information on all Focal Press publications
visit our website at www.focalpress.com

Contents

Preface

Television has come a long way in its short history, and the accompanying sound has also made great leaps. So many, in fact, that sound for television has become a sufficiently expansive subject to warrant a book of its own. The rate of change in technology has accelerated recently, with the introduction first of stereo then multi-channel or surround sound with television. The onward march of digital techniques first embraced production and then moved on to consumer devices and program delivery.

The rapid change has made it difficult for those working in this field to keep up with the technology and even harder for those just setting out on a career in television sound. This book is designed to make life a little easier by outlining all of the relevant principles and practice.

As in all of my books I have kept to plain English in preference to mathematics wherever possible. Facts are not stated, but are derived from explanations which makes learning the subject much easier. This is the only viable approach when the range of disciplines to be covered is so wide.

It is not so long ago that digital audio was considered a specialist subject. This is no longer the case; it has moved into the mainstream of audio recording, production and distribution. Consequently the approach taken in this volume is to consider analog and digital audio as alternatives and to stress the advantages of both. Digital audio is only audio with numbers and someone who understands analog audio well will find digital audio much more approachable.

Audio is only as good as the transducers employed, and consequently microphone and loudspeaker technology features heavily here. It has been a pleasure to explode a few myths about loudspeakers.

The newcomer will be able to handle the initial level of presentation, whereas the more experienced reader will treat this as a refresher and gain more value from the explanations of new technology. I have learned that those who believe education and entertainment have nothing in common probably know little about either. Consequently I have not let the advanced technology in this volume completely exclude light-heartedness.

John Watkinson

Chapter 1

Sound and hearing

In this chapter the characteristics of sound as an airborne vibration and as a human sensation are tied together.

1.1 What is sound?

There is a well-known philosophical riddle which goes 'If a tree falls in the forest and no one is there to hear it, does it make a sound?' This question can have a number of answers depending on the plane one chooses to consider. I believe that to understand what sound really is requires us to interpret this on many planes. Physics can tell us the mechanism by which the sound propagates through the air, and biology can tell us how the ear works. Psychoacoustics can describe how much of what is heard is actually perceived. I believe we are still struggling to explain why we enjoy music and why certain sounds can make us happy and others can reduce us to tears.

Television viewing is more than information transfer. Almost without exception, television programmes, and indeed commercials, are designed to appeal to the emotional side of human nature. It must be stressed that sound is an essential component of that experience, just as it is in the cinema. So let us not forget, whatever the technology we deal with, there is a common goal of delivering a satisfying experience to the viewer. And we cannot neglect our own satisfaction. Possibly because it is a challenging and multidisciplinary subject, doing a good job in audio is particularly rewarding.

1.2 The physics of sound

Sound is simply an airborne version of vibration, which is why the two topics are inextricably linked. The air which carries sound is a mixture of gases, mostly nitrogen, some oxygen, a little carbon dioxide and so on. Gases are the highest energy state of matter, which is another way of saying that you have to heat ice to get water then heat it some more to get steam. The reason that a gas takes up so much more room than a liquid is that the molecules contain so much energy that they break free from their neighbours and rush around at high speed. As Figure 1.1(a) shows, the innumerable elastic collisions of these high speed molecules produce pressure on the walls of any gas container. In fact the distance a

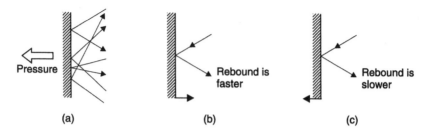

Figure 1.1 (a) The pressure exerted by a gas is due to countless elastic collisions between gas molecules and the walls of the container. (b) If the wall moves against the gas pressure, the rebound velocity increases. (c) Motion with the gas pressure reduces the particle velocity.

molecule can go without a collision, the mean free path, is quite short at atmospheric pressure. Consequently gas molecules also collide with each other elastically, so that if left undisturbed, in a container at a constant temperature, every molecule would end up with essentially the same energy and the pressure throughout would be constant and uniform.

Sound disturbs this simple picture. Figure 1.1(b) shows that a solid object which moves *against* gas pressure increases the velocity of the rebounding molecules, whereas in Figure 1.1(c) one moving *with* gas pressure reduces that velocity. The average velocity of all of the molecules in a layer of air near to a moving body is the same as the velocity of the body. This results in a local increase or decrease in pressure. Thus sound is both a pressure and a velocity disturbance. Integration of the velocity disturbance gives the displacement. This disturbance is not transmitted instantly throughout the gas. The disturbed molecules communicate energy to other molecules via collisions and this takes time because of the mean free path. The rate at which a disturbance propagates in this way is called the speed of sound, c. The disturbance can be due to a one-off event known as percussion, or a periodic event such as the sinusoidal vibration of a tuning fork. The sound due to percussion is called transient, whereas a periodic stimulus produces steady-state sound having a pitch or frequency f. As sound consists of pressure and velocity changes, either of these can be measured. Pressure is a directionless or *scalar* quantity, consequently pressure microphones give an output independent of the direction from which the sound approached. Velocity, however, is a *vector* quantity having an associated direction. Velocity measurement will give a result which depends on direction. If the velocity is measured in a direction at right angles to the direction of propagation of a sound, it will always be zero.

The sensation of sound is proportional to the velocity. However, the displacement is the integral of the velocity. Figure 1.2 shows that to obtain an identical velocity or slope the amplitude must increase as the inverse of the frequency. Consequently for a given sound pressure level (SPL) low frequency sounds result in much larger air movement than high frequency. The SPL is proportional to the *volume velocity* of the source which is obtained by multiplying the vibrating area by the velocity of movement.

The speed of sound is independent of pressure, however rapid pressure changes will change the temperature.

If the temperature is raised, the molecules move faster and so they collide more often and the speed of sound goes up. The speed of sound is proportional

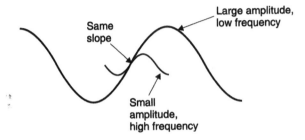

Same
slope

Large amplitude,
low frequency

Small
amplitude,
high frequency

Figure 1.2 For a given velocity or slope, lower frequencies require greater amplitude.

to the square root of the absolute temperature. Atmospheric pressure at sea level is reasonably constant, with the weather causing changes of around 1%. Temperature changes with respect to absolute zero (–273°C) also amount to around 1% except in extremely inhospitable places. The speed of sound experienced by most of us is about 1000 feet per second or 344 metres per second.

The speed of sound is not independent of frequency. High frequencies travel slightly faster than low. Figure 1.3 shows that a complex sound source produces harmonics whose phase relationship with the fundamental advances with the distance the sound propagates. This allows one mechanism (there are others) by which one can judge the distance from a known sound source. Clearly for realistic sound reproduction nothing in the audio chain must distort the phase relationship between frequencies. A system which accurately preserves such relationships is said to be phase linear.

Because sound travels at a finite speed, the fixed observer at some distance from the source will experience the disturbance at some later time. In the case of a transient, the observer will detect a single replica of the original as it passes at the speed of sound. In the case of the tuning fork, a periodic sound, the pressure peaks and dips follow one another away from the source at the speed of sound. For a given rate of vibration of the source, a given peak will have propagated a constant distance before the next peak occurs. This distance is called the wavelength, (λ). Figure 1.4 shows that wavelength is defined as the distance between any two identical points on the whole cycle. If the source vibrates faster, successive peaks get closer together and the wavelength gets shorter. Figure 1.4 also shows that the wavelength is inversely proportional to the frequency. It is easy to remember that the wavelength of 1000 Hz is a foot (about 30 cm).

Near source   At a distance

Figure 1.3 In a complex waveform, high frequencies travel slightly faster producing a relative phase change with distance.

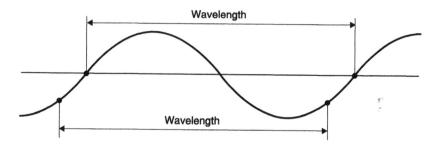

Figure 1.4 Wavelength is defined as the distance between two points at the same place on adjacent cycles. Wavelength is inversely proportional to frequency.

If there is relative motion between the source and the observer, the frequency of a periodic sound will be changed. Figure 1.5 shows a sound source moving towards the observer. At the end of a cycle, the source will be nearer the observer than at the beginning of the cycle. As a result the wavelength radiated in the direction of the observer will be shortened so that the pitch rises. The wavelength of sounds radiated away from the observer will be lengthened. The same effect will occur if the observer moves. This is the Doppler effect which is most noticeable on passing motor vehicles whose engine notes appear to drop as they pass. Note that the effect always occurs, but it is only noticeable on a periodic sound. Where the sound is aperiodic, such as broadband noise, the Doppler shift will not be heard.

The speed of sound is a function of rapid pressure changes, yet sound is a pressure variation. One might expect some effects because of this. Fortunately, sounds which are below the threshold of pain have such a small pressure variation compared with atmospheric pressure that the effect is negligible and air can be assumed linear. However, on any occasion where the pressures are higher, this is not a valid assumption. In such cases the positive half cycle significantly

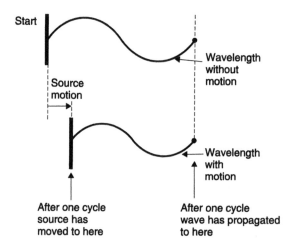

Figure 1.5 Periodic sounds are subject to Doppler shift if there is relative motion between the source and the observer.

increases ambient pressure and the speed of sound, whereas the negative half cycle reduces pressure and velocity. Figure 1.6 shows that this results in significant distortion of a sine wave, ultimately causing a *shock wave* which can travel faster than the speed of sound until the pressure has dissipated with distance. This effect is responsible for the sharp sound of a handclap.

1.3 The ear

By definition, the sound quality of an audio system can only be assessed by human hearing. Many items of audio equipment can only be designed well with a good knowledge of the human hearing mechanism. The acuity of the human ear is astonishing. It can detect tiny amounts of distortion, and will accept an enormous dynamic range over a wide number of octaves. If the ear detects a different degree of impairment between two audio systems in properly conducted tests, we can say that one of them is superior. Thus quality is completely subjective and can only be checked by listening tests. However, any characteristic of a signal which can be heard can in principle also be measured by a suitable instrument. The subjective tests will tell us how sensitive the instrument should be. Then the objective readings from the instrument give an indication of how acceptable a signal is in respect of that characteristic.

The sense we call hearing results from acoustic, mechanical, nervous and mental processes in the ear/brain combination, leading to the term psychoacoustics. It is only possible briefly to introduce the subject here. The interested reader is also referred to Moore [1.1] for an excellent treatment.

Figure 1.7 shows that the structure of the ear is traditionally divided into the outer, middle and inner ears. The outer ear works at low impedance, the inner ear works at high impedance, and the middle ear is an impedance matching device. The visible part of the outer ear is called the pinna, which plays a subtle role in determining the direction of arrival of sound at high frequencies. It is too small to have any effect at low frequencies. Incident sound enters the auditory canal or meatus and vibrates the eardrum or tympanic membrane. The inner ear or cochlea works by sound travelling though a fluid. Sound enters the cochlea via a membrane called the oval window. If airborne sound were to be incident

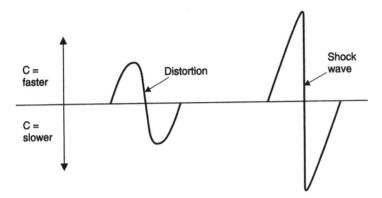

Figure 1.6 At high level, sound distorts itself by increasing the speed of propagation on positive half cycles. The result is a shock wave or N-wave.

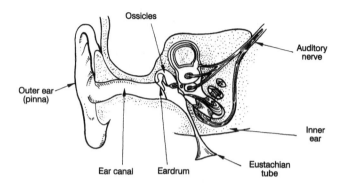

Figure 1.7 The structure of the human ear. See text for details.

on the oval window directly, the serious impedance mismatch would cause most of the sound to be reflected. The middle ear remedies that mismatch by providing a mechanical advantage. The tympanic membrane is linked to the oval window by three bones known as ossicles which act as a lever system such that a large displacement of the tympanic membrane results in a smaller displacement of the oval window but with greater force. The area of the tympanic membrane is greater than that of the oval window, again multiplying the available force. Consequently small pressures over the large area of the tympanic membrane are converted to high pressures over the small area of the oval window.

The three bones, the malleus, the incus and the stapes, are located by minute muscles which are normally relaxed. However, the middle ear reflex is an involuntary tightening of the muscles which heavily damps their ability to transmit sound at low frequencies. The main function of this reflex is to reduce the audibility of one's own speech.

The cochlea is a spiral cavity within bony walls which is filled with fluid. The cochlea is divided lengthwise by the basilar membrane. Vibrations from the stapes are transferred to the oval window and become fluid pressure variations. These act upon the basilar membrane which is not uniform, but which tapers in width and varies in thickness. The part of the basilar membrane which resonates as a result of an applied sound is a function of the frequency. High frequencies cause resonance near to the oval window, whereas low frequencies cause resonances further away. The basilar membrane is active in that it contains elements which can generate vibration as well as to sense it. These are connected in a regenerative fashion so that the Q factor, or frequency selectivity, is higher than it would otherwise be. The high frequencies are detected at the end of the membrane nearest to the eardrum and the low frequencies are detected at the opposite end. The ear analyses with frequency bands, known as critical bands, about 100 Hz wide below 500 Hz and from one-sixth to one-third of an octave wide, proportional to frequency, above this. Critical bands were first described by Fletcher [1.2]. Later Zwicker experimentally established 24 critical bands [1.3], but Moore and Glasberg suggested narrower bands [1.4].

In the presence of a complex spectrum, the ear fails to register energy in some bands when there is more energy in a nearby band. The vibration of the membrane in sympathy with a single frequency cannot be localized to an infinitely small area, and nearby areas are forced to vibrate at the same frequency with an amplitude that decreases with distance. Within those areas, other frequencies are excluded unless the amplitude is high enough to dominate the local vibration of the membrane. Thus the Q factor of the membrane is responsible for the degree of auditory masking, defined as the decreased audibility of one sound in the presence of another.

The degree of masking depends upon whether the masking tone is a sinusoid, which gives the least masking, or noise [1.5]. However, harmonic distortion produces widely spaced frequencies, and these are easily detected even in minute quantities by a part of the basilar membrane which is distant from that part which is responding to the fundamental. The sensitivity of the ear to distortion probably deserves more attention in audio equipment than the fidelity of the dynamic range or frequency response. The masking effect is asymmetrically disposed around the masking frequency [1.6]. Above the masking frequency, masking is more pronounced, and its extent increases with acoustic level. Below the masking frequency, the extent of masking drops sharply at as much as 90 dB per octave.

Owing to the resonant nature of the membrane, it cannot start or stop vibrating rapidly. The spectrum sensed changes slowly even if that of the original sound does not. The reduction in information sent to the brain is considerable; masking can take place even when the masking tone begins after and ceases before the masked sound. This is referred to as forward and backward masking [1.7]. An example of the slowness of the ear is the Haas effect, in which the direction from which a sound is perceived to have come is determined from the first arriving wavefront. Later echoes simply increase the perceived loudness, as they have the same spectrum and increase the existing excitation of the membrane. Far from being a defect in human hearing, the Haas effect is a distinct advantage because it allows increased intelligibility in reverberant surroundings. Our simple microphones have no such ability, which is why we often need to have acoustic treatment in areas where microphones are used.

At its best, the ear can detect a sound pressure variation of only 2×10^{-5} pascals (Pa) r.m.s. and so this figure is used as the reference against which SPL is measured. The sensation of loudness is a logarithmic function of SPL and consequently a logarithmic unit, the decibel, (dB) is used in audio measurement. The decibel is explained in detail in Section 3.2.

The dynamic range of the ear exceeds 130 dB, but at the extremes of this range, the ear is either straining to hear or is in pain. Neither of these cases can be described as pleasurable or entertaining, and it is hardly necessary to produce audio of this dynamic range in television since, among other things, the viewer is unlikely to have anywhere sufficiently quiet to listen or equipment of sufficient power to reproduce the loudest events. On the other hand, extended listening to music whose dynamic range has been excessively compressed is fatiguing.

The frequency response of the ear is not at all uniform and it also changes with SPL. Figure 1.8 shows the actual SPLs judged to be equally loud as a given level at 1 kHz. The most dramatic effect is that the bass content of reproduced sound is disproportionately reduced as the level is turned down. This would suggest that if a powerful yet high quality reproduction system is available, the

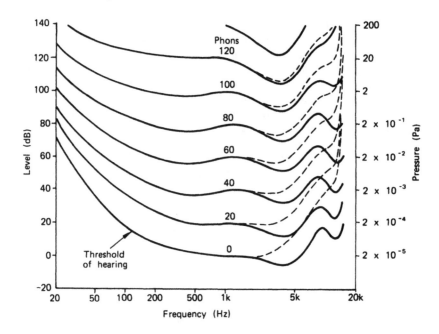

Figure 1.8 Contours of equal loudness showing that the frequency response of the ear is highly level dependent (solid line, age 20; dashed line, age 60).

correct tonal balance can be obtained simply by setting the volume control to the correct level. This is indeed the case. Audio systems with a more modest specification would have to resort to a bass-boost tone control to achieve a correct tonal balance at lower SPL.

A further consequence is that recordings which are mixed at an excessively high level will appear bass light when played back at a normal level.

Usually, people's ears are at their most sensitive between about 2 kHz and 5 kHz, and although some people can detect 20 kHz at high level, there is much evidence to suggest that most listeners cannot tell if the upper frequency limit of sound is 20 kHz or 16 kHz [1.8, 1.9]. For a long time it was thought that frequencies below about 40 Hz were unimportant, but it is becoming clear that reproduction of frequencies down to 20 Hz improves reality and ambience [1.10]. The generally accepted frequency range for high quality audio is 20 Hz to 20 000 Hz, although for broadcasting, an upper limit of 15 000 Hz is often applied.

1.4 Sound propagation

The frequency range of human hearing is extremely wide, covering some ten octaves (an octave is a doubling of pitch or frequency) without interruption. There is hardly any other engineering discipline in which such a wide range is found. For example in radio different wavebands are used so that the octave span of each is quite small. Whilst video signals have a wide octave span, the

signal-to-noise and distortion criteria for video are extremely modest in comparison. Consequently audio is one of the most challenging subjects in engineering.
Whilst the octave span required by audio can be met in electronic equipment, the
design of mechanical transducers such as microphones and loudspeakers will
always be difficult.

Sound is a wave motion, and the way a wave interacts with any object
depends upon the relative size of that object and the wavelength. The audible
range of wavelengths is from around 17 millimetres to 17 metres, so dramatic
changes in the behaviour of sound over the frequency range should be expected.

Figure 1.9(a) shows that when the wavelength of sound is large compared to
the size of a solid body, the sound will pass around it almost as if it were not
there. When the object is large compared to the wavelength, then simple reflection
takes place, as in Figure 1.9(b). However, when the size of the object and the
wavelength are comparable, the result can only be explained by diffraction theory.

The parameter which is used to describe this change of behaviour with frequency is known as the wave number k and is defined as:

$$k = \frac{2\pi f}{c} = \frac{2\pi}{\lambda}$$

where f = frequency, c = the speed of sound and λ = wavelength. In practice the
size of any object a in metres is multiplied by k.

A good rule of thumb is that below $ka = 1$, sound tends to pass around as in
Figure 1.9(a) whereas above $ka = 1$, sound tends to reflect as in Figure 1.9(b).

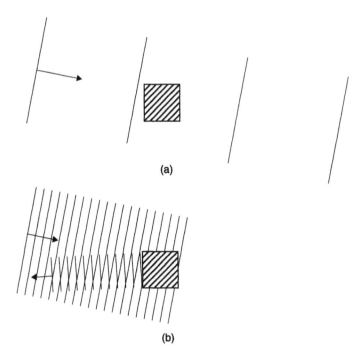

(a)

(b)

Figure 1.9 (a) Sound waves whose spacing is large compared to an obstacle simply pass round it. (b) When
the relative size is reversed, an obstacle becomes a reflector.

Figure 1.10 shows that when two sounds of equal amplitude and frequency add together, the result is completely dependent on the relative phase of the two. At (a), when the phases are identical, the result is the arithmetic sum. At (b), where there is a 180 degree relationship, the result is complete cancellation. This is constructive and destructive interference. At any other phase and/or amplitude relationship, the result can only be obtained by vector addition as shown in (c).

The wave theory of propagation of sound is based on interference and suggests that a wavefront advances because an infinite number of point sources can be considered to emit spherical waves which will only add when they are all in the same phase. This can only occur in the plane of the wavefront. Figure 1.11(a) shows that at all other angles, interference between spherical waves is destructive. For any radiating body, such as a vibrating object, it is easy to see from Figure 1.11(b) that when ka is small, only weak spherical radiation is possible, whereas when ka is large, a directional plane wave can be propagated or beamed.

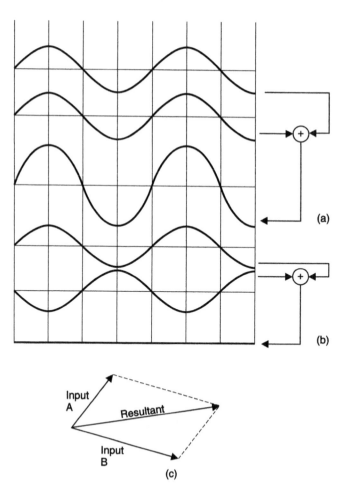

Figure 1.10 (a) Constructive interference between two in-phase signals. (b) Destructive interference between out-of-phase signals. (c) Vector addition is needed to find result of arbitrary phase relationship.

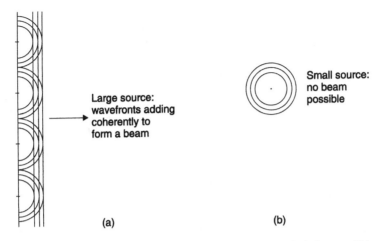

Figure 1.11 (a) Plane waves can be considered to propagate as an infinity of spherical waves which cancel out in all directions other than forward to form a beam. (b) Where the sound source is small no beam can be formed.

Consequently high frequency sound behaves far more directionally than low frequency sound. As SPL is proportional to volume velocity, as frequency falls the volume or displacement must rise. This means that low frequency sound can only be radiated effectively by large objects, hence all of the bass instruments in the orchestra are much larger than their treble equivalents. This is also the reason why a loudspeaker cone is only seen to move at low frequencies.

When a wavefront arrives at a solid body, it can be considered that the surface of the body acts as an infinite number of points which re-radiate the incident sound in all directions. It will be seen that when ka is large and the surface is flat, constructive interference only occurs when the wavefront is *reflected* such that the angle of reflection is the same as the angle of incidence. When ka is small, the amount of re-radiation from the body compared to the radiation in the wavefront is very small. Constructive interference takes place beyond the body as if it were absent, thus it is correct to say that the sound diffracts around the body.

If sound enters a medium in which the speed is different, the wavelength will change causing the wavefront to leave the interface at a different angle. This is known as refraction. The ratio of velocity in air to velocity in the medium is known as the refractive index of that medium; it determines the relationship between the angles of the incident and refracted wavefronts. This does not happen much in real life; it requires a thin membrane with different gases each side to demonstrate the effect. However, as was shown above in connection with the Doppler effect, wind has the ability to change the wavelength of sound. Figure 1.12 shows that when there is a wind blowing, friction with the earth's surface causes a velocity gradient. Sound radiated upwind will have its wavelength shortened more away from the ground than near it, whereas the reverse occurs downwind. Thus upwind it is difficult to hear a sound source because the radiation has been refracted upwards, whereas downwind the radiation will be refracted towards the ground, making the sound 'carry' better. Temperature gradients can have the same effect, giving the acoustic equivalent of a mirage.

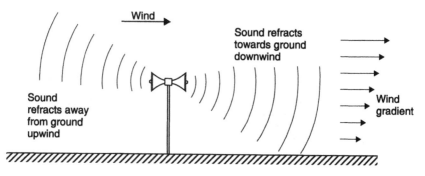

Figure 1.12 When there is a wind, the velocity gradient refracts sound downwards downwind of the source and upwards upwind of the source.

When two sounds of equal frequency and amplitude are travelling in opposite directions, the result is a *standing wave* where constructive interference occurs at fixed points one wavelength apart with nulls between. This effect can often be found between parallel hard walls, where the space will contain a whole number of wavelengths. As Figure 1.13 shows, a variety of different frequencies can excite standing waves at a given spacing. Wind instruments work on the principle of standing waves. The wind produces broadband noise, and the instrument resonates at the fundamental depending on the length of the pipe. The higher harmonics add to the richness or *timbre* of the sound produced.

In practice, many real materials do not reflect sound perfectly. As Figure 1.14 shows, some sound is reflected, some is transmitted and the remainder is absorbed. The proportions of each will generally vary with frequency. Only porous materials are capable of being effective sound absorbers. The air movement is slowed by viscous friction among the fibres. Such materials include wood, foam, cloth and carpet. Non-porous materials either reflect or transmit according to their mass. Thin, hard materials such as glass, reflect high frequencies but transmit low frequencies. Substantial mass is required to prevent transmission of low frequencies, there being no substitute for masonry.

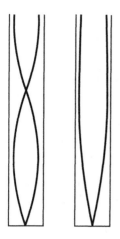

Figure 1.13 Standing waves in an organ pipe can exist at several different frequencies.

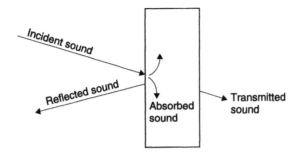

Figure 1.14 Incident sound is partially reflected, partially transmitted and partially absorbed. The proportions vary from one material to another and with frequency.

In real rooms with hard walls, standing waves can be set up in many dimensions, as Figure 1.15 shows. The frequencies at which the dominant standing waves occur are called eigentones. Any sound produced in such a room which coincides in frequency with an eigentone will be strongly emphasized as a resonance which might take some time to decay. Clearly a cube would be the worst possible shape for a studio as it would have a small number of very powerful resonances.

At the opposite extreme, an *anechoic chamber* is a room treated with efficient absorption on every surface. Figure 1.16 shows that long wedges of foam absorb sound by repeated reflection and absorption down to a frequency determined by the length of the wedges (our friend *ka* again). Some people become distressed in anechoic rooms and musical instruments sound quiet, lifeless and boring. Sound of this kind is described as *dry*.

Reflected sound is needed in concert halls to amplify the instruments and add richness or *reverberation* to the sound. Since reflection cannot and should not be eliminated, in practice studios, listening rooms and concert halls are designed so that resonances are made as numerous and close together as possible, in order that no single one appears dominant. Apart from choosing an irregular shape, this

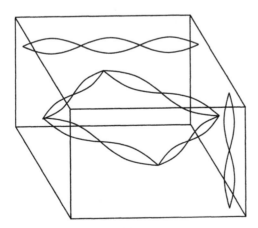

Figure 1.15 In a room, standing waves can be set up in three dimensions.

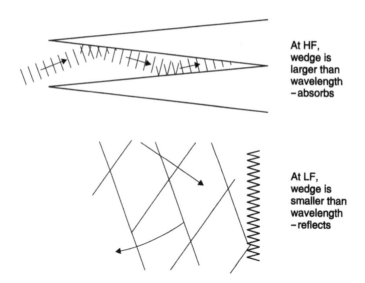

At HF,
wedge is
larger than
wavelength
– absorbs

At LF,
wedge is
smaller than
wavelength
– reflects

Figure 1.16 Anechoic wedges are effective until wavelength becomes too large to see them.

goal can be helped by the use of *diffusers* which are highly irregular reflectors. Figure 1.17 shows that if a two-plane stepped surface is made from a reflecting material, at some wavelengths there will be destructive interference between sound reflected from the upper surface and sound reflected from the lower. Consequently the sound cannot reflect back the way it came but must diffract off at any angle where constructive interference can occur. A diffuser made with steps of various dimensions will reflect sound in a complex manner. Diffusers are thus very good at preventing standing waves without the deadening effect that absorbing the sound would have.

In a hall having highly reflective walls, any sound will continue to reflect around for some time after the source has ceased. Clearly as more absorbent is introduced, this time will fall. The time taken for the sound to decay by 60 dB is known as the *reverberation time* of the room. The optimum reverberation time depends upon the kind of use to which the hall is put. Long reverberation times

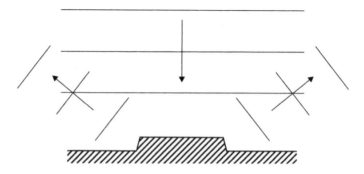

Figure 1.17 A diffuser prevents simple reflection of an incident wavefront by destructive interference. The diffracted sound must leave by another path.

make orchestral music sound rich and full, but would result in intelligibility loss on speech. Consequently theatres and cinemas have short reverberation times, opera houses have medium times and concert halls have the longest. In some multi-purpose halls the reverberation can be modified by rotating wall panelling, although more recently this is done with electronic artificial reverberation using microphones, signal processors and loudspeakers.

Only porous materials make effective absorbers, but these cannot be used in areas which are prone to dampness or where frequent cleaning is required. This is why indoor swimming pools are so noisy.

1.5 Bass tip-up

A transducer can be affected dramatically by the presence of other objects, but the effect is highly frequency dependent. In Figure 1.18(a) a high frequency is radiated, and this simply reflects from the nearby object because the wavelength is short and the object is acoustically distant or in the *far field*. However, if the wavelength is made longer than the distance between the source and the object as in Figure 1.18(b), the object is acoustically close or in the *near field* and becomes

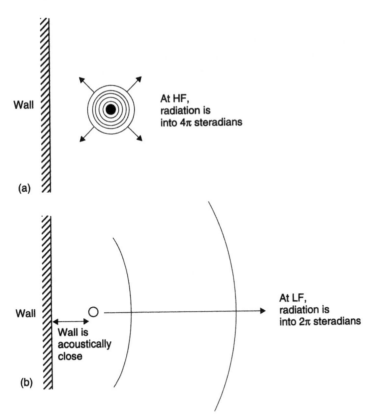

Figure 1.18 (a) At high frequencies an object is in the far field. (b) At low frequencies the same object is in the near field and increases velocity by constricting the radiation path.

part of the source. The effect is that the object reduces the solid angle into which radiation can take place as well as raising the acoustic impedance the transducer sees. The volume velocity of the source is confined into a smaller cross-sectional area and consequently the velocity must rise. This is the phenomenon of bass tip-up, which has a number of manifestations in practical audio equipment.

In Figure 1.19 the effect of positioning a loudspeaker is shown. In free space (a) the speaker might show a reduction in low frequencies which disappears when it is placed on the floor (b). In this case placing the speaker too close to a wall, or even worse, in a corner, (c), will emphasize the low frequency output. High quality loudspeakers will have an adjustment to compensate for positioning. The technique can be useful in the case of small cheap loudspeakers whose low frequency (LF) response is generally inadequate. Some improvement can be made by corner mounting.

The effect has to be taken into account when stereo loudspeakers are installed. At low frequencies the two speakers will be acoustically close and so will mutually raise their acoustic impedance causing a potential bass tip-up problem. When a pair of stereo speakers has been properly equalized, disconnecting one will result in the remaining speaker sounding bass light.

When close to a point source the particle velocity rises disproportionately because it is not in phase with the pressure. This causes the *proximity effect*, which clearly affects velocity microphones more than it does pressure microphones.

In Figure 1.20 the effect of positioning a microphone too close to a source is shown. This is most noticeable in public address systems where the gain is limited to avoid howl-round. The microphone must then be held close to obtain

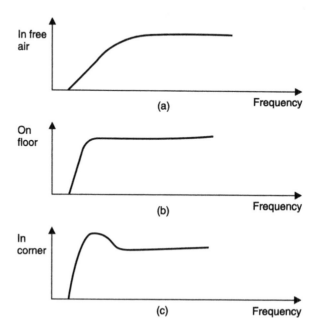

Figure 1.19 Loudspeaker positioning affects LF response. (a) Speaker in free air appears bass deficient. (b) This effect disappears when floor mounted. (c) Bass is increased when mounted near a wall or corner.

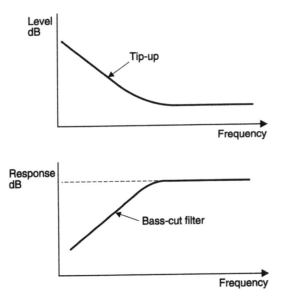

Figure 1.20 Bass tip-up due to close microphone positioning. A suitable filter will help intelligibility.

sufficient level and the plosive parts of speech are emphasized, cutting intelligibility. The solution is an LF cut filter which de-emphasizes the microphone output at low frequency. Some microphones incorporate a switchable filter. In recording, where there is no possibility of howl-round, the correct solution is to place the microphone at a greater distance.

In Figure 1.21 a supra-aural headphone (one which sits above the ear rather than over it) in free space has a very poor LF response because it is a dipole source and at low frequency air simply moves from front to back in a short circuit. However, the presence of the listener's head obstructs the short circuit and the bass tip-up effect gives a beneficial extension of frequency response to the intended listener, whilst those not wearing the headphones only hear high frequencies. Many personal stereo players incorporate an LF boost to equalize the losses further. All practical headphones must be designed to take account of the presence of the user's head since headphones work primarily in the near field.

A dramatic example of the near field effect is obtained by bringing the ear close to the edge of a cymbal shortly after it has been struck. The fundamental note which may only be a few tens of Hz can clearly be heard. As the cymbal is such a poor radiator at this frequency there is very little damping of the fundamental, which will continue for some time. At normal distances it is quite inaudible.

1.6 Directionality in hearing

We can determine reasonably well where a sound is coming from, but how is this done? An understanding of the mechanisms of direction sensing is important for the successful implementation of spatial illusions such as stereophonic sound.

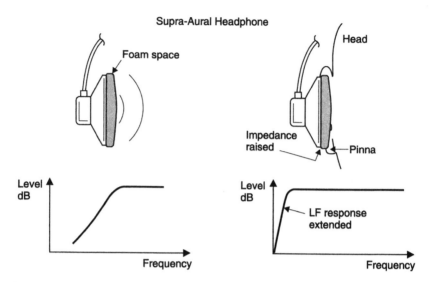

Figure 1.21 Supra-aural headphones rely on the bass tip-up in the near field to give a reasonable bass response.

Ownership of a pair of spaced ears is a good start. As Figure 1.22 shows, this allows a number of mechanisms. At (a) a phase shift will be apparent between the two versions of a tone picked up the two ears unless the source of the tone is dead ahead (or behind). At (b) the distant ear is shaded by the head, resulting in reduced response compared to the nearer ear. At (c) a transient sound arrives later at the more distant ear.

If the phase shift mechanism (a) is considered, then it will be clear that there will be considerable variation in this effect with frequency. At a low frequency such as 30 Hz, the wavelength is around 11.5 metres. Even if heard from the side, the ear spacing of about 0.2 metres will result in a phase shift of only 6 degrees and so this mechanism must be quite poor at low frequencies.

At a high frequency such as 10 kHz, the ear spacing is many wavelengths, and variations in the path length difference will produce a confusing and complex phase relationship. Consequently the phase comparison mechanism must be restricted to frequencies where the wavelength is short enough to give a reasonable phase shift, but not so short that complete cycles of shift are introduced. In fact this restriction corresponds quite well to the range of frequencies in human speech which carry the intelligibility.

The shading mechanism (b) will be dominated by our friend *ka*, suggesting that at low and middle frequencies sound will diffract round the head sufficiently well that there will be no significant difference between the level at the two ears. Only at high frequencies does sound become directional enough for the head to shade the distant ear. At very high frequencies, the shape of the pinnae must have some effect on the sound, which is a function of direction. It is thought that the pinnae allow some height discrimination.

The problem with tones or single frequencies is that they produce a sinusoidal waveform, one cycle of which looks much like another leading to ambiguities in the time between two versions. This is shown in Figure 1.23(a). Pure tones are

extremely difficult to localize, especially as they will often excite room resonances which give a completely misleading impression of the location of the
sound source.

Fortunately real world sounds often contain transients, especially those
sounds which indicate a potential hazard. Transients differ from tones in that they
contain many different frequencies. A transient also has an unique aperiodic

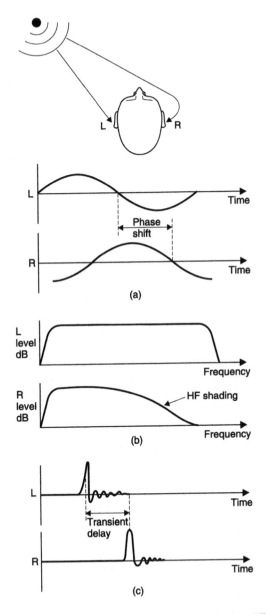

Figure 1.22 Having two spaced ears is cool. (a) Off-centre sounds result in phase difference. (b) Distant ear
is shaded by head producing loss of HF. (c) Distant ear detects transient later.

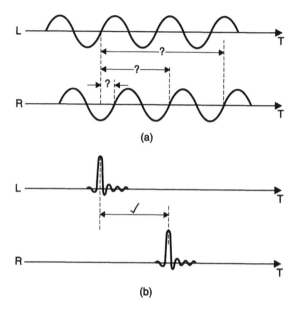

Figure 1.23 (a) Pure tones cause ambiguity in timing differences. (b) Transients have no ambiguity and are easier to localize.

waveform which Figure 1.23(b) shows has the advantage that there can be no ambiguity in the time delay between two versions.

The basilar membrane has been described as a frequency analysis device which produces nerve impulses according to which frequencies are present in the incident sound. Clearly when a transient arrives from one side, several frequencies will be excited simultaneously in the nearer ear, closely followed by a similar excitation in the farther ear. It will be relatively easy for the brain to determine the time between these large similar excitations, so although phase differences and shading play a part, the most effective lateral direction sensing mechanism is analysis of the arrival time of transients. This is easily demonstrated by wearing a blindfold and having a helper move around the room making a variety of noises. The helper will be easier to locate when making clicking noises than when humming.

As the ears are almost exactly at each side of the head it is quite possible for sound sources ahead or behind to produce the same relative delay, phase shift and shading, resulting in an ambiguity. There are two main ways in which the ambiguity can be resolved. If a plausible source of sound can be seen, then clearly the source is not behind. In real life, visual clues often override auditory clues allowing for some inaccuracy in the spatial realism of the sound when accompanied by a picture. However, if the inaccuracy is too great fatigue may result.

A second way of resolving front/back ambiguity is to turn the head slightly. This is often done involuntarily and most people are not aware they are using the technique. In fact when people deliberately try harder to locate a sound they often keep their head quite still making the ambiguity worse.

Laterally separated ears are ideal for determining the location of sound sources in the plane of the earth's surface, which is after all where most sound sources emanate. Our ability to determine height in sound is very poor. People who look up when they hear birdsong may not be able to determine the height of the source at all, they may simply know, as we all do, that birds sing in trees.

Figure 1.24 shows that when standing, sounds from above reach the ear directly and via a ground reflection which has come via a longer path (there is also a smaller effect due to reflection from the shoulders). At certain frequencies the extra path length will correspond to a 180-degree phase shift, causing cancellation at the ear. The result is a frequency response consisting of evenly spaced nulls which is called *comb filtering*. A moving object such as a plane flying over will suffer changing geometry which will cause the frequency of the nulls to fall towards the point where the overhead position is reached. The result depends on the kind of aircraft. In a piston engined aircraft, the sound is dominated by discrete frequencies such as the exhaust note. When such a discrete spectrum is passed through a swept comb filter the result is simply that the levels of the various components rise and fall. The Doppler effect has the major effect on the pitch which appears to drop suddenly as the plane passes. In the case of a jet plane, the noise emitted on approach is broad band white noise from the compressor turbines. As stated above, the Doppler effect has no effect on aperiodic noise. When passed through a swept comb filter, the pitch of the output appears to fall giving the characteristic descending whistle of a passing jet.

The direction sensing ability has been examined by making binaural recordings using miniature microphones actually placed down the ear canals of a volunteer. When these are played back on headphones to the person whose ears were used for the recording, full localization of direction including front/rear and height discrimination is obtained. However, the differences between people's ears are such that the results of playing the recording to someone else are much worse. The same result is obtained if a dummy head is used.

Whilst binaural recordings give very realistic spatial effects, these effects are only obtained on headphones and consequently the technique is unsatisfactory for signals intended for loudspeaker reproduction and is not used in television.

When considering the localization ability of sound, it should be appreciated that vision can produce a very strong clue. If only one person can be seen speaking in a crowd, then any speech heard must be coming from that person. The same is true when watching films or television. This is a bonus because it means

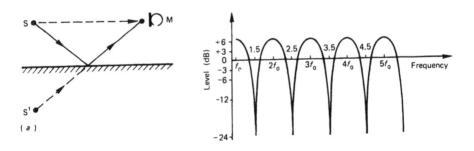

Figure 1.24 Comb filtering effect produced by interference between a sound wave and a delayed version of itself. (a) Microphone M receives direct wave from source S and a delayed wave from image sound source S¹. (b) Frequency response shows alternating peaks and troughs resembling the teeth of a comb.

Chapter 2

Microphones, loudspeakers and stereophony

In this chapter the essential topics of microphones and loudspeakers are considered. The polar or directional characteristics of transducers are their most important attribute and are treated here before consideration of the operating principle. It will be seen that polar characteristics assume even greater importance if the illusion of stereophony is to be made realistic. The microphone is a measuring device and its output consists of *intelligence* rather than power. Consequently it is possible to conceive of an ideal microphone and the best practice approaches this quite closely. The same is not true in loudspeakers because of the requirement to transfer power.

2.1 Introduction

We use electrical signalling to carry audio because it allows ease of processing and simple transmission over great distances using cable or radio. It is essential to have transducers such as microphones and loudspeakers which can convert between real sound and an electrical equivalent. Figure 2.1(a) shows that even if the electrical system is ideal, the overall quality of a sound system is limited by the quality of both microphone and loudspeaker. Figure 2.1(b) shows that in a television sound system, the quality of the final loudspeaker is variable, dependent upon what the viewer can afford. However, it must be assumed that at least a number of viewers will have high quality systems. Consequently the microphone used in production must be of high quality and the loudspeakers used for monitoring the production process must also be of high quality so that the quality of the broadcast sound exceeds or at least equals that of the viewer's equipment, even after all of the recording, production and distribution processes have been carried out.

2.2 Microphone principles

The job of the microphone is to convert sound into an electrical signal. As was seen in Section 1.2, sound consists of both pressure and velocity variations, and microphones can use either or both in order to obtain various directional characteristics.

Figure 2.2(a) shows a true pressure microphone which consists of a diaphragm stretched across an otherwise sealed chamber. In practice a small

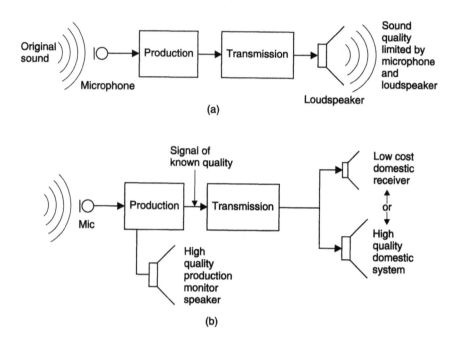

Figure 2.1 (a) The final sound quality of an audio system is limited by both microphones and loudspeakers. (b) Sound production must be performed using high quality loudspeakers on the assumption that the transmitted quality should be limited only by the viewer's equipment.

pinhole is provided to allow changes in atmospheric pressure to take place without causing damage. Some means is provided to sense the diaphragm motion and convert it into an electrical output signal. This can be done in several ways which will be considered in Section 2.4. The output of such a microphone for small values of ka is completely independent of direction as Figure 2.2(b) shows.

Figure 2.2(d) shows a true velocity microphone, also called a *pressure gradient* microphone, in which the diaphragm is suspended in free air from a perimeter frame. The maximum excursion of the diaphragm will occur when it faces squarely across the incident sound. As Figure 2.2(e) shows, the output will fall as the sound moves away from this axis, reaching a null at 90 degrees. If the diaphragm were truly weightless then it would follow the variations in air velocity perfectly. However as the diaphragm has finite mass then a pressure difference is required to make it move, hence the term pressure gradient microphone which is strictly more accurate.

In practice the directional characteristics shown in Figure 2.2(b) and (e) are redrawn in *polar coordinates* such that the magnitude of the response of the microphone corresponds to the distance from the centre point at any angle. The pressure microphone (c) has a circular polar diagram as it is omnidirectional or *omni* for short. Omni microphones are good at picking up ambience and reverberation, which makes them attractive for music and sound effects recordings in good locations. In acoustically poor locations they cannot be used because they are unable to discriminate between wanted and unwanted sound. Directional microphones are used instead.

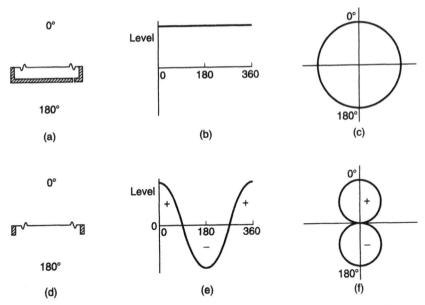

Figure 2.2 (a) Pressure microphone only allows sound to reach one side of the diaphragm. (b) Pressure microphone is omnidirectional for small ka. (c) Directional characteristic is more intuitive when displayed in polar form. (d) Velocity or pressure gradient microphone exposes both sides of diaphragm. (e) Output of velocity microphone is a sinusoidal function of direction. (f) In polar coordinates velocity microphone shows characteristic figure-of-eight shape for small ka.

The velocity microphone has a polar diagram (f) which is the shape of a figure-of-eight. Note the null at 90 degrees and that the polarity of the output reverses beyond 90 degrees giving rise to the term *dipole*. The figure-of-eight microphone (sometimes just called 'an *eight*') responds in two directions giving a degree of ambience pickup, although the sound will be a little drier than that of an omni. A great advantage of the figure-of-eight microphone over the omni is that it can reject an unwanted sound. Rather than point the microphone at the wanted sound, a better result will be obtained by pointing the null or dip in the polar diagram at the source of the unwanted sound.

Unfortunately the velocity microphone cannot distinguish between velocity variations due to sound and those due to gusts of wind. Consequently velocity microphones are more sensitive to wind noise than omnis.

If an omni and an eight are mounted coincidentally, various useful results can be obtained by combining the outputs. Figure 2.3(a) shows that if the omni and eight signals are added equally, the result is a heart-shaped polar diagram called a *cardioid*. This response is obtained because at the back of the eight the output is antiphase and has to be subtracted from the output of the omni. With equal signals this results in a null at the rear and a doubling at the front. This useful polar response will naturally sound drier than an eight, but will have the advantage of rejecting more unwanted sound under poor conditions. In public address applications, use of a cardioid will help to prevent *feedback* or *howl-round* which occurs when the microphone picks up too much of the signal from the loudspeakers. Virtually all hand-held microphones have a cardioid response where the major lobe faces axially so that the microphone is pointed *at* the sound source. This is known as an *end-fire* configuration, shown in Figure 2.3(b). Where only a cardioid response is required, this can be obtained using a single diaphragm where the chamber behind it is not sealed, but open to the air via an acoustic labyrinth which gives some resistance to sound to obtain a omni trend, but which also allows ambient sound to reach both sides of the diaphragm to allow a velocity characteristic. The sum of the two effects gives the wanted cardioid response.

In more flexible microphones there are two diaphragms, and the relative amounts of signal from the two diaphragms can be varied by a control. By disabling one signal, a cardioid response can be obtained. Combining them equally results in an omni or an eight, depending on the phase relationship. Combining them unequally results in a sub-cardioid shown in Figure 2.3(c) or a hyper-cardioid shown in Figure 2.3(d). Where a variable polar response is required, the end-fire configuration cannot be used as the microphone body would then block the rearward access to the diaphragm. The *side-fire* configuration is shown in Figure 2.3(e), where the microphone is positioned *across* the approaching sound, usually in a vertical position. For television applications where the microphone has to be out of shot, such microphones are often slung from above pointing vertically downwards.

In most applications the polar diagrams noted above are adequate, but on occasions it proves to be quite impossible to approach the subject close enough

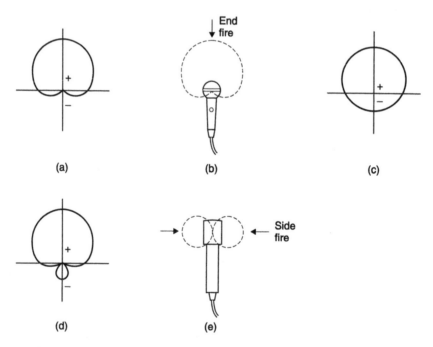

(a)   (b)   (c)

(d)   (e)

Figure 2.3 (a) Combining an omni response with that of an eight in equal amounts produces the useful cardioid directivity pattern. (b) Hand-held fixed cardioid response microphones are usually built in the end-fire configuration where the body is placed in the null. (c) Sub-cardioid obtained by having more omni in the mix gives better ambience pickup than cardioid. (d) Hyper-cardioid obtained by having more eight in the mix is more directional than cardioid but the presence of the rear lobe must be considered in practice. (e) Microphones with variable polar diagram are generally built in the side-fire configuration.

and then a highly directional microphone is needed. Figure 2.4(a) shows that the *shotgun* microphone consists of a conventional microphone capsule which is mounted at one end of a slotted tube. Sound wavefronts approaching from an angle will be diffracted by the slots such that each slot becomes a re-radiator launching sound down the inside of the tube. However, Figure 2.4(b) shows that the radiation from the slots travelling down the tube will not add coherently and will be largely cancelled. A wavefront approaching directly on axis as in (c) will pass directly down the outside and the inside of the tube as if the tube were not there and consequently will give a maximum output.

2.3 Microphone limitations

The directivity patterns given in Section 2.2 are only true where ka is small and are thus ideal. In practice at high frequencies ka will not be small and the actual polar diagram will differ due to diffraction becoming significant.

Figure 2.5(a) shows the result of a high frequency sound arriving off-axis at a large diaphragm. It will be clear that at different parts of the diaphragm the sound has a different phase and that in an extreme case cancellation will occur, reducing the output significantly.

When the sound is even further off-axis, shading will occur. Consequently at high frequency the polar diagram of a nominally omni microphone may look something like that shown in Figure 2.5(b) and that of an eight may resemble Figure 2.5(c). Note the narrowing of the response such that proper reproduction of high frequencies is only achieved when the source is close to the axis.

It is possible to reduce ka by making the microphone diaphragm smaller but this results in smaller signals making low noise difficult to achieve. However, developments in low noise circuitry will allow diaphragm size beneficially to be reduced.

In the case of the shotgun microphone, the tube will become acoustically small at low frequencies and will become ineffective. The polar diagram will widen. As such microphones are normally used for speech, the addition of a high pass filter removes low frequencies without affecting the speech quality.

Unlike human hearing, which is selective, microphones reproduce every sound which reaches them. Figure 2.6(a) shows the result of placing a micro-

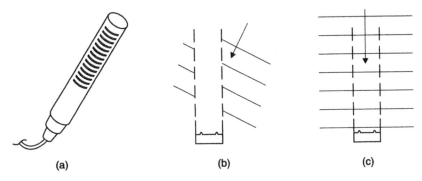

(a)   (b)   (c)

Figure 2.4 (a) Structure of a shotgun microphone consists of a slotted tube with the capsule at one end. (b) Off-axis sounds produce an incoherent diffraction pattern in the tube due to multi-path cancellation. (c) On-axis wavefront is not aware of the presence of the tube.

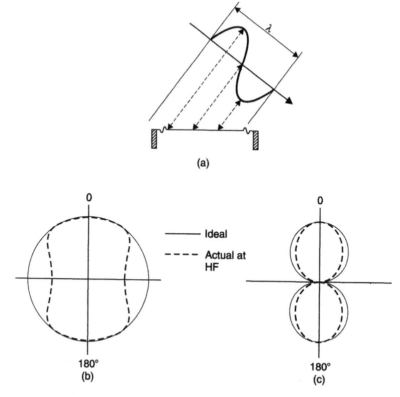

(a)

(b)

(c)

Figure 2.5 (a) Off-axis response is impaired when *ka* is not small because the wavefront reaches different parts of diaphragm at different times causing an aperture effect. (b) Polar diagram of practical omni microphone at high frequency shows narrowing of frontal response due to aperture effect and rear loss due to shading. (c) Practical eight microphone has narrowing response at high frequency.

phone near to a hard wall. The microphone receives a combination of direct and reflected sound between which there is a path length difference. At frequencies where this amounts to a multiple of a wavelength, the reflection will reinforce the direct sound, but at intermediate frequencies cancellation will occur, giving a comb filtering effect. Clearly a conventional microphone should not be positioned near a reflecting object.

The path length difference is zero at the wall itself. The pressure zone microphone (Figure 2.6(b)) is designed to be placed on flat surfaces where it will not suffer from reflections.

2.4 Microphone mechanisms

There are two basic mechanisms upon which microphone operation is based: electrostatic and electrodynamic. The former is generally considered to be of better quality, but more expensive and delicate, whereas the latter is considered to be of inferior quality but cheaper and more robust.

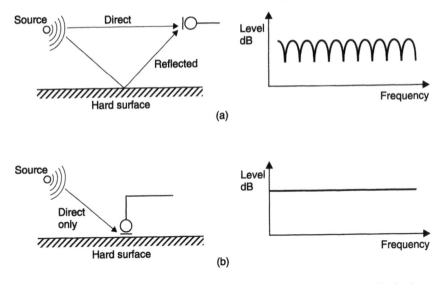

Figure 2.6 (a) Microphone placed several wavelengths from reflective object suffers comb filtering due to path length difference. (b) Pressure zone microphone is designed to be placed at a boundary where there is no path length difference.

The electrodynamic microphone operates on the principle that a conductor moving through a magnetic field will generate a voltage proportional to the rate of change of flux. As the magnetic flux is constant, then this results in an output proportional to velocity.

There are two basic implementations of the electrodynamic microphone: the ribbon and the moving coil. Figure 2.7(a) shows that in the ribbon microphone the diaphragm is a very light metallic foil which is suspended between the poles of a powerful magnet. The incident sound causes the diaphragm to move and the velocity of the motion results in an EMF being generated across the ends of the ribbon. The most common form which the ribbon microphone takes is the figure-of-eight response. The output voltage of the ribbon is very small but the source impedance is very low and so it is possible to use a transformer to produce a higher output voltage at a more convenient impedance.

The advantage of the ribbon microphone is that the motion of the ribbon is directly converted to an electrical signal. This is potentially very accurate. However, unless the transformer is of extremely high quality the inherent accuracy will be lost. A further problem is that to obtain reasonable sensitivity the diaphragm must be relatively large, leading to the directivity problems mentioned in Section 2.3. The magnet must also be large, leading to a heavy construction. A further problem is that the diaphragm is extremely delicate and a single exposure to wind might destroy it. Although the ribbon microphone was at one time the best available, it has effectively been made obsolete by the capacitor microphone and is little used today.

The most common version of the electrodynamic microphone is the moving coil system shown in Figure 2.7(b). The diaphragm is connected to a cylindrical former upon which is wound a light coil of wire. The coil operates in the radial flux pattern of a cylindrical magnet. As it is possible to wind many turns of wire

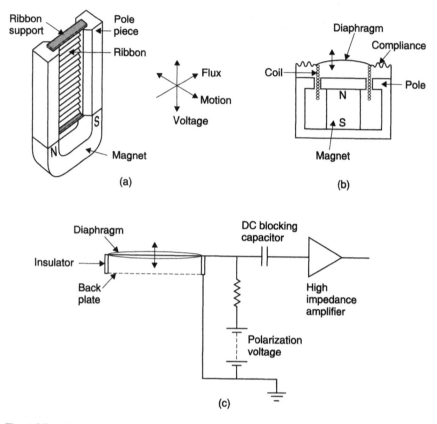

Figure 2.7 (a) Now obsolete ribbon microphone suspends conductive diaphragm in magnetic field. Low output impedance requires matching transformer. (b) The moving coil microphone is robust but indirect coupling impairs quality. (c) Capacitor microphone has very tight coupling but requires high impedance electronics and needs to be kept dry.

on the coil, the output of such a microphone is relatively high. The structure is quite robust and can easily withstand wind and poor handling. However, the indirect conversion, whereby the sound moves the diaphragm and the diaphragm moves the coil, gives impaired performance because the coil increases the moving mass and the mechanical coupling between the coil and diaphragm is never ideal. Consequently the moving coil microphone, generally in a cardioid response form, finds common application in outdoor use for speech or for public address, but is considered inadequate for accurate music work.

The electrostatic microphone works on the variation in capacitance between a moving diaphragm and a fixed plate. As the diaphragm needs no other mechanism there is extremely direct coupling between the sound waveform and the electrical output and so very high quality can be achieved. There are two forms of electrostatic microphone: the condenser, or capacitor, microphone and the electret microphone.

In the condenser microphone the diaphragm is highly insulated from the body of the microphone and is fed with a high polarizing voltage via a large resistance. Figure 2.7(c) shows that a fixed metallic grid forms a capacitor in conjunction with the diaphragm. The grid is connected to an amplifier having a very high impedance. The high impedances on both plates of the capacitor mean that there is essentially a constant charge condition. Consequently when incident sound moves the diaphragm and the capacitance between it and the grid varies, the result will be a change of voltage at the grid which can be amplified to produce an output.

The condenser microphone requires active circuitry close to the capsule and this requires a source of DC power. This is often provided using the same wires as the audio output, adopting the principle of *phantom powering* described in Section 3.6.

If the impedance seen by the grid is not extremely high, charge can leak away when the diaphragm moves. This will result in poor output at low frequencies. As the condenser microphone requires high impedances to work properly, it is at a disadvantage in damp conditions which means that in practice it has to be kept indoors in all but the most favourable weather. Some condenser microphones contain a heating element which is designed to drive out moisture. In older designs based on vacuum tubes, the heat from the tube filaments would serve the same purpose. If a capacitor microphone has become damp, it may create a great deal of output noise until it has dried out.

In the electret microphone a crystalline material is employed which can produce a constant electric field without power. A diaphragm moving in such a field will produce an output which will usually require to be locally amplified. Whilst phantom power can be used, electret microphones are often powered by a small dry cell incorporated into the microphone body.

2.5 Loudspeaker concepts

Whilst a microphone can in principle and in practice produce a very accurate measurement of the sound field *approaching* the point where it is located, the loudspeaker is not so fortunate. Loudspeakers usually approximate to a point source of sound and produce a sound field *leaving* that point. Clearly a single loudspeaker is producing sound travelling in exactly the opposite direction to the original. Figure 2.8 shows the problem. Sound approaching a microphone at (a) does so from a multiplicity of sources, whereas sound leaving a single loudspeaker superimposes all of these sources into one. Consequently a monophonic or single loudspeaker is doomed to heap one sound source upon another and if heard in anechoic conditions (b) this is exactly what happens. Whilst the waveform might be reproduced with great precision, the spatial characteristics of such a sound are very disappointing. However, when listening in a room having a degree of reverberation a better result is achieved. The human listener is accustomed to ambient sound approaching from all directions in real life and when this does not happen in a reproduction system the result is unsatisfactory. To avoid this in all real listening environments a considerable amount of reverberant sound is required in addition to the direct sound from the loudspeakers. Figure 2.8(c) shows that the reverberation of the listening room results in sound approaching the listener from all sides, giving a closer approximation to situation in (a).

Better spatial accuracy requires more channels and more loudspeakers. Whilst the ideal requires an infinite number of loudspeakers, it will be seen that, with

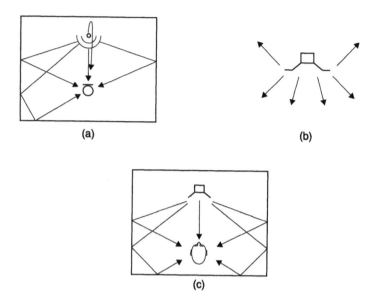

Figure 2.8 (a) Sound approaches microphone from many directions due to ambiance and reverberation. (b) In anechoic conditions single loudspeaker produces exactly the opposite of (a). (c) Loudspeaker in reverberant conditions simulates situation of (a) at listener's ears.

care, as few as two speakers can give a convincing spatial illusion. Two speakers can only give spatial accuracy for sound sources located between them. Reverberation in the listening room then provides ambient sound from all remaining directions. Clearly the resultant reverberant sound field can never be a replica of that at the microphone, but a plausible substitute is essential for realism.

If such realism is to be achieved, the polar diagram of the loudspeaker and its stability with frequency is extremely important. A common shortcoming with loudspeakers is that output becomes more directional with increasing frequency. Figure 2.9(a) shows that although the frequency response on-axis may be ruler flat, giving a good quality direct sound, the frequency response off-axis may be quite badly impaired as at (b). However, the off-axis output is necessary to excite the essential reverberant field and as a result the tonal balance of the reverberation will not match that of the direct sound [2.1].

The resultant conflict may only be perceived subconsciously and cause *listening fatigue* where the initial impression of the loudspeaker is quite good, but after a while one starts looking for excuses to stop listening. The hallmark of a good loudspeaker installation is that one can listen to it indefinitely and that of an excellent installation is where one does not want to stop.

Unfortunately such instances are rare. More often loudspeakers are used having such poor polar characteristics that the only remedy is to make the room highly absorbent so that the direct sound dominates. This has led to the well-established myth that reflections are bad and that extensive room treatment is necessary for good monitoring. The problem is compounded by the fact that an absorbent room requires more sound power to obtain a given SPL. Consequently heavily treated rooms require high power loudspeakers which often further sacrifice polar response in order to achieve that high power.

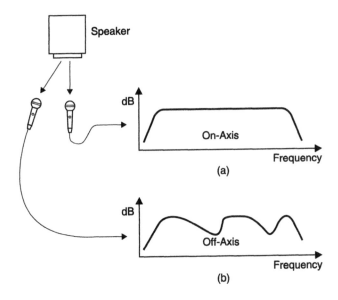

Figure 2.9 (a) Ideal on-axis response is achieved by many loudspeakers. (b) Off-axis response of most loudspeakers is irregular causing colouration of reverberant field.

One need not understand much about diffraction to appreciate that a conventional box-shaped loudspeaker with drive units in the front will suffer extensive shading of the radiation to the rear and thus will create a coloured reverberant field. Clearly a much more effective way of exciting reverberation with an accurate tonal balance is for the loudspeaker to emit sound to the rear as well as to the front. This is the advantage of the dipole loudspeaker which has a figure-of-eight polar diagram.

2.6 Practical loudspeakers

In all practical loudspeakers some form of diaphragm has to be vibrated which then vibrates the air [2.2]. There are contradictory requirements. As was seen in Section 1.4 the SPL which the loudspeaker can generate is determined by the volume velocity. If the frequency is halved, the displacement must be doubled either by doubling the area of the diaphragm or by doubling the travel or some combination of both. Figure 2.10 shows the requirements for various conditions. Clearly a powerful loudspeaker which is able to reproduce the lowest audio frequencies must have a large diaphragm capable of considerable travel. Unfortunately any diaphragm which is adequate for low frequency use will have an extremely large figure of ka at high frequencies and Section 1.4 showed that this would result in *beaming* or high directivity, which is undesirable.

One solution is to use a number of drive units which each handle only part of the frequency range. Those producing low frequencies will have large diaphragms with considerable travel, whereas those producing high frequencies will have small diaphragms whose movement is seldom visible. A frequency dividing system or *crossover network* is required to limit the frequency range of signals to each drive unit.

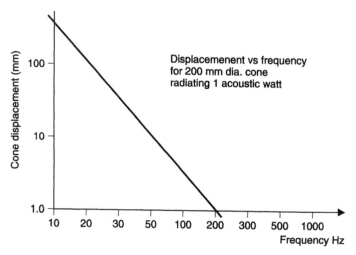

Figure 2.10 Volume velocity requirements dictate that low frequency reproduction must involve large diaphragms or long travel.

The two transduction mechanisms used in microphones are both reversible and so can also be applied to loudspeakers. The electrodynamic loudspeaker produces a force by passing current through a magnetic field, whereas the electrostatic loudspeaker produces force due to the action of an electric field upon a charge.

As with the microphone, the electrodynamic loudspeaker can be made using a ribbon or a moving coil. The ribbon loudspeaker is identical in construction to the ribbon microphone shown in Figure 2.7. Only the transformer needs to be larger in order to handle appreciable power. The ribbon is not capable of much volume velocity and so is usually restricted to working at high frequencies.

The moving coil loudspeaker [2.3] is by far the most common device. Figure 2.11(a) shows the structure of a typical low cost unit containing an annular ferrite magnet. The magnet produces a radial field in which the coil operates. The coil drives the centre of the diaphragm which is supported by a *spider*, allowing axial but not radial movement. The perimeter of the cone is supported by a flexible *surround*.

A major drawback of the moving coil loudspeaker is that the drive force is concentrated in the centre of the diaphragm, whereas the air load is distributed over the surface. This can cause the diaphragm, and the sound, to be distorted. This distortion can be minimized in an LF unit because extra mass is tolerable and the cone can be stiffer.

The low cost ferrite magnet is a source of considerable problems in the television environment because it produces so much stray flux. This can cause colour purity errors and distortion of the picture on a video monitor because it disturbs the magnetic deflection system. Whilst screening can be installed, a better solution for speakers to be used in television is to employ a different magnetic circuit design as shown in Figure 2.11(b). Such a design completely contains its own flux because the magnet is inside the magnetic circuit. The magnet has to be smaller, but sufficient flux is easily available using rare-earth magnets.

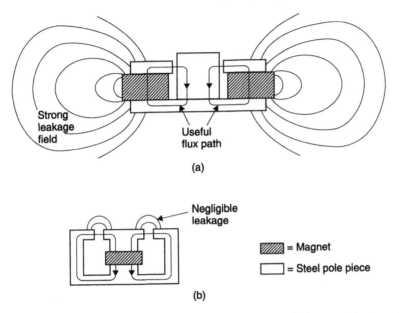

Strong leakage field

Useful flux path

(a)

Negligible leakage

▨ = Magnet

☐ = Steel pole piece

(b)

Figure 2.11 (a) The conventional moving coil loudspeaker with a low cost ferrite magnet showing serious flux leakage. Using a higher energy rare earth magnet the magnetic circuit can be completely self-contained as in (b).

The electrostatic loudspeaker is shown in Figure 2.12. A slightly conductive diaphragm is connected to a high voltage DC supply so that it becomes charged. The high resistivity of the diaphragm prevents the charge moving around so that at all audio frequencies it can be considered fixed. Any charge placed in an electric field will experience a force. The electric field is provided by electrodes either side of the diaphragm which are driven in antiphase, often by a centre-tapped transformer.

The advantage of the electrostatic loudspeaker is that the driving mechanism is fundamentally linear and the mechanical drive is applied uniformly all over the diaphragm. Consequently there is no reason why the diaphragm should distort.

As the density of air is so low, the mass of air a loudspeaker actually moves is a few percent of the mass of the diaphragm. Consequently most of the drive power supplied to any kind of loudspeaker is wasted driving the diaphragm back and forth and the efficiency is very poor. There is no prospect of this situation ever being resolved. Where high SPL is required, a horn may be used to couple the diaphragm more efficiently to the air load by acting as an acoustic transformer. Figure 2.13 shows that a horn may be used with a ribbon or a moving coil driver.

Unfortunately the horn has a number of drawbacks. Unless large compared to the wavelength, the horn has no effect. Consequently the horn is restricted to mid-range frequencies and above. A further problem is that the mouth of the horn forms an impedance mismatch and standing waves occur down the horn, causing the frequency response to be irregular. Perhaps the greatest drawback of the horn is that the SPL at the diaphragm is considerably higher than at the mouth. Where sound pressures become an appreciable proportion of atmospheric pressure, the

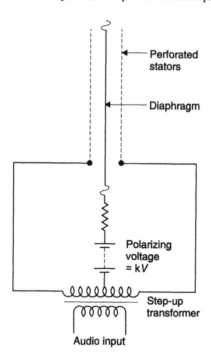

Figure 2.12 The electrostatic loudspeaker uses the force experienced by a charge in an electric field. The charge is obtained by polarizing the diaphragm.

effect noted in Section 1.2 occurs in which the air becomes non-linear and distortion occurs. Consequently for high quality applications the horn is ruled out.

One of the greatest challenges in a loudspeaker is to make the polar characteristics change smoothly with frequency in order to give an uncoloured reverberant field. Unfortunately crossing over between a number of drive units often does not achieve this. Figure 2.14 shows that at the crossover frequency both drive units separated by a are contributing equally to the radiation. If ka is small the system acts as a single driver and the polar diagram will be undisturbed. However with typical values of a this will only be true below a few hundred hertz. If a

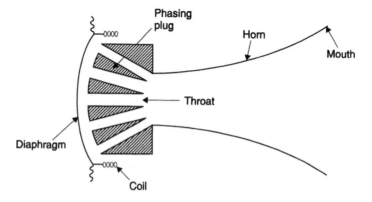

Figure 2.13 Horn loading may be used to improve the efficiency of moving coil or ribbon drivers. Horns can suffer from standing waves and distortion at high throat SPL.

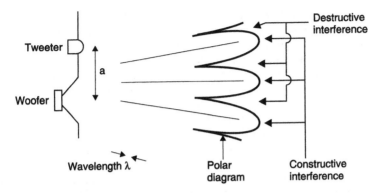

Figure 2.14 At the crossover frequency both drivers are operating and if *ka* is not small the polar diagram becomes highly irregular because of path length differences.

crossover is attempted above that frequency a diffraction pattern will be created where the radiation will sum or cancel according to the path length differences between the two drivers. This results in an irregular polar diagram and some quite undesirable off-axis frequency responses [2.4].

Clearly the traditional loudspeaker with many different drive units is inadequate. Certain moving coil and electrostatic transducers can approach the ideal with sufficient care. Figure 2.15(a) shows that if the flare angle of a cone-type moving coil unit is correct for the material, the forward component of the speed of sound in the cone can be made slightly less than the speed of sound in the air,

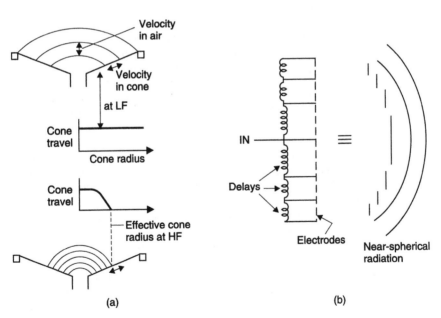

(a)   (b)

Figure 2.15 (a) A cone built as a lossy transmission line can reduce its diameter as a function of frequency, giving smooth directivity characteristics. (b) Using delay lines a flat electrostatic panel can be made to behave like a pulsating sphere.

so that nearly spherical wavefronts can be launched. The cone is acting as a mechanical transmission line for vibrations which start at the coil former and work outwards. If frequency dependent loss is introduced into the transmission line, the higher the frequency the smaller is the area of the cone which radiates. Done correctly the result is a constant dispersion drive unit. There are vibrations travelling out across the cone surface and the cone surround must act as a matched terminator so that there can be no reflections.

The elegant solution in an electrostatic speaker first proposed by Walker of Quad [2.5] is to make the mechanically flat diaphragm behave like a sphere by splitting the electrode structure into concentric rings fed by lossy delay lines as shown in Figure 2.15(b). This produces what is known as a *phased array*. The outward propagation of vibrations across the diaphragm again simulates quite closely a sector of a pulsating sphere. Again matched termination at the perimeter is required to prevent reflections.

Using either of these techniques allows the construction of a single drive unit which will work over the entire mid and treble range and display smooth directivity changes. A low ka crossover to a low frequency driver completes the design. Such two-way speakers can display extremely good performance especially if implemented with active techniques.

Interestingly enough if strict polar response and distortion criteria are applied, the phased array electrostatic loudspeaker turns out to be capable of higher SPL than the moving coil unit. This is because the phased array approach allows the electrostatic loudspeaker to have a very large area diaphragm without beaming taking place. Consequently at mid and high frequencies it can achieve very large volume velocities. Unfortunately space can seldom be found in television applications for a large panel speaker.

At low values of ka air will simply flow from one side of a diaphragm to the other without much radiation taking place. Consequently effective LF loudspeakers require the rear of the diaphragm to be enclosed. The air thus enclosed is compressed as the diaphragm moves back and acts as an air spring whose stiffness increases as the box is made smaller. The stiffness of the air spring is added to the stiffness of the drive unit suspension to obtain the total stiffness. The mass of the diaphragm and the total stiffness determines the *fundamental resonance* of the loudspeaker.

The light diaphragm of the electrostatic loudspeaker results in excessively high fundamental resonance if placed in an enclosure. The short travel allowed by the fixed electrodes also precludes operation at LF. Consequently the electrostatic principle is best employed as a mid and high frequency dipole in conjunction with a low distortion moving coil LF unit.

Rice and Kellog discovered around 1925 that the amplitude of motion of a moving coil loudspeaker cone reaches a peak at the resonance, and falls at 6 dB per octave either side of that resonance as shown in Figure 2.16(a). Radiation is proportional to cone *velocity* which is obtained by differentiating the position. Differentiation tilts the response by 6 dB/octave. Consequently as Figure 2.16(b) shows, the radiation is independent of frequency above resonance but falls at 12 dB/octave below. Figure 2.16(c) shows that with an amplifier of low output resistance, the mechanical resonance is damped by the coil resistance. The moving coil motor acts as a transformer coupling the resonant system to the damping resistance. Increasing the flux density or the length of coil in the gap increases the effective ratio of the transformer and makes the coil resistance appear lower, increasing the damping. The peakiness of the resonance is adjusted in the design process by balancing the coil resistance, the strength of the magnet and the length of wire in the coil. If the drive unit is not driven by a low output impedance via low resistance cables the resonance may be under-damped and a pronounced peak or *honk* may occur.

To obtain reproduction of lowest frequencies, the resonance must be kept low and this implies a large box to reduce the stiffness of the air spring.

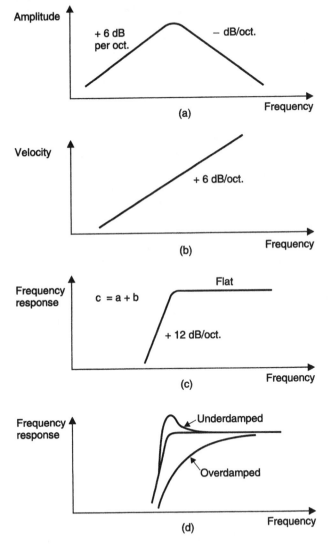

Figure 2.16 (a) The amplitude of a resonant system falls at 6 dB/octave away from the peak. (b) The velocity of the system in (a) is obtained by differentiating the displacement, resulting in a 6 dB/octave tilt. This gives a flat response region with a 12 dB/octave roll-off below resonance. (c) Moving coil motor acts as a transformer coupling resonant system to the coil resistance which acts as a damper. Motor design affects peakiness of resonance (d).

Unfortunately a large box forms a very poor place from which to radiate high frequencies. Figure 2.17 shows that when an HF unit is fitted in a large box, diffraction results in re-radiation at the corners [2.6, 2.7]. When the direct and re-radiated sounds combine the result is constructive or destructive interference depending on the frequency. This causes ripples in the on-axis frequency response. The larger the box the further down the spectrum these ripples go. The effect can be reduced by making the cabinet smaller and making it curved with no sharp corners.

In early attempts to give satisfactory LF performance from smaller boxes, a number of passive schemes have been tried. These include the reflex cabinet shown in Figure 2.18(a) which has a port containing an air mass. This is designed to resonate with the air spring at a frequency below that of the fundamental resonance of the driver so that as the driver response falls off the port output takes over. In some designs the air mass is replaced by a compliantly mounted diaphragm having no coil, known as an ABR or auxiliary bass radiator, (b).

Another alternative is the transmission line speaker shown in (c), in which the rear wave from the driver is passed down a long damped labyrinth which emerges at a port. The length is designed to introduce a 180 degree phase shift at the frequency where the port output is meant to augment the driver output. A true transmission line loudspeaker is quite large in order to make the labyrinth long enough. Some smaller models are available which claim to work on the transmission line principle, but in fact the labyrinth is far too short and there is a chamber behind the drive unit which makes these heavily damped reflex cabinets.

More recently the *bandpass* enclosure (d) has become popular, probably because suitable computer programs are now available to assist the otherwise difficult design calculations. The bandpass enclosure has two chambers with the drive unit between them. All radiation is via the port.

The reflex, ABR, bandpass and transmission line principles have numerous drawbacks the most serious of which are that the principle only works on continuous tone. Low frequency transients suffer badly from linear distortion because

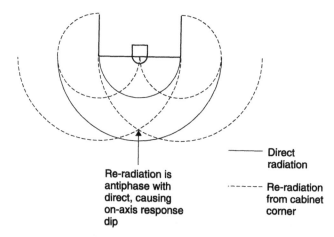

Re-radiation is
antiphase with
direct, causing
on-axis response
dip

——— Direct
radiation

- - - - - Re-radiation
from cabinet
corner

Figure 2.17 Attempting to radiate HF from a tweeter mounted in a large rectangular box produces frequency response irregularity due to diffraction from the box corners.

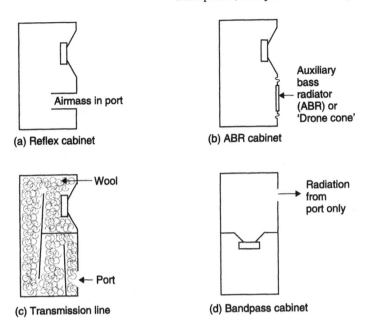

Figure 2.18 Various attempts to reproduce low frequencies. (a) Mass of air in reflex duct resonates with air spring in box. (b) Air mass replaced by undriven diaphragm or ABR. (c) Rear wave is phase shifted 180° in transmission line to augment front radiation. (d) Bandpass enclosure puts drive unit between two resonating chambers. None of these techniques can properly reproduce transients and active techniques have rendered them obsolete.

the leading edge of the transients are removed and reproduced after the signal has finished, to give the phenomenon of *hangover*. Different order filtering and different damping change the amount by which this takes place, but can never eliminate the problem which is most noticeable on transient musical information such as percussion and on effects such as door slams. It is quite impossible to use such an LF technique in conjunction with an electrostatic HF unit or a constant directivity mid-top moving coil unit because the quality mismatch is too obvious.

2.7 Active loudspeakers

With modern electronic techniques all of these passive schemes must be considered obsolete. An active loudspeaker containing its own amplifiers can easily introduce equalization and signal processing which can artificially move the fundamental resonance down in frequency and achieve any desired damping factor. The advantage of this approach is that the speaker can be made relatively phase linear and will not suffer from hangover. The smaller cabinet also allows the radiation characteristics of the mid and high frequencies to be improved. The sensitivities of the drive units do not need to be matched so that there is more freedom in their design. In the absence of separate amplifier cabinets less space is needed overall and the dubious merits of exotic speaker cables become irrelevant.

Figure 2.19 shows the block diagram of a modern active loudspeaker. The line-level input passes to a crossover filter which routes low and high frequencies to each driver. Each drive unit has its own power amplifier. The low frequency power amplifier is preceded by a compensation network which can electronically lower the resonance of the LF driver and determine the damping factor. In some units the LF diaphragm is fitted with a feedback transducer so that distortion can be further reduced.

Audio power amplifiers have a relatively simple task in that they simply have to make an audio signal bigger. Unfortunately they also have to be reasonably efficient. Figure 2.20 shows some approaches to power amplifiers. The Class A amplifier has potentially the lowest distortion because the entire waveform is handled by the same devices. The high standing current required results in a serious heat dissipation problem which deters all but enthusiasts. The Class B amplifier has different devices to provide the positive and negative halves of the cycle. This reduces dissipation but it is extremely difficult to prevent crossover distortion taking place at low signal levels where the handover from one polarity to the other takes place. The Class D or switching amplifier is extremely efficient because its output devices are either on or off. A variable output is obtained by controlling the duty cycle of the switching and filtering the output. Whilst switching amplifiers cannot produce low enough distortion for high quality mid and treble drive, they offer an ideal solution to efficient generation of low frequencies. Such amplifiers save on cost and weight because they can use a smaller power supply and need no massive heatsinks.

In most amplifiers linearity is achieved using negative feedback which compares the output with the input to create a correction signal. In most semiconductor amplifiers the open loop characteristics are not very good and a lot of feedback is necessary, requiring a lot of open loop gain which is not always available at all frequencies. When a fast slewing input arrives, the amplifier may go open loop and produce transient intermodulation distortion (TIM).

In the *error correcting* amplifier topology the output of a small Class A high performance amplifier is added to that from a powerful but average Class B amplifier. The drive to the high power amplifier is arranged so that it always delivers enough current to the load to prevent the small amplifier from clipping.

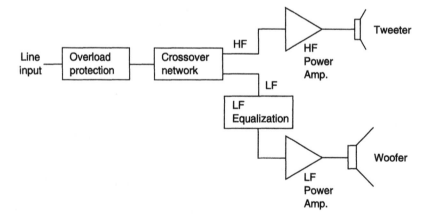

Figure 2.19 In an active loudspeaker the crossover function is performed at line level and separate amplifiers are used for each drive unit. The LF response is controlled by equalization and electronic damping allowing accurate transient reproduction.

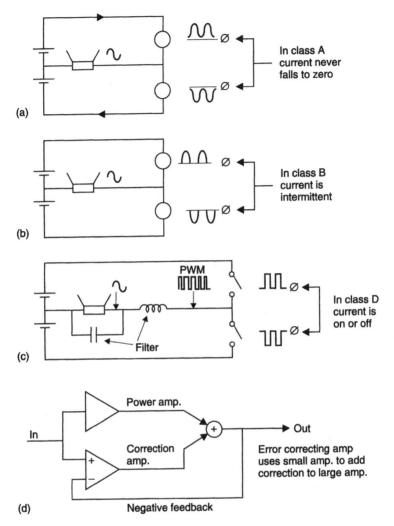

Figure 2.20 (a) Class A amplifier has low distortion but heavy standing current produces heat problem. (b) Class B amplifier is more efficient but introduces crossover distortion. (c) Class D amplifier is very efficient and ideal for driving active woofers. (d) Error correcting amplifier uses small, wide bandwidth Class A amplifier to compensate for distortions of Class B stage.

Careful feedback arrangements allow the small amplifier to cancel the distortions of the large one. The low distortion of a Class A amplifier can then be combined with the reduced heat dissipation of a Class B amplifier.

With very few exceptions, audio amplifiers are voltage sources with a very low output impedance. No attempt is made to impedance match the loudspeaker to the amplifier in the conventional sense as there is no transmission line involved. A lower impedance loudspeaker simply takes more power for a given amplifier output voltage. When driven to the extreme an amplifier will run out of supply voltage if driving a high impedance, whereas it will current limit if driving a low impedance. There is an optimum load impedance into which the amplifier can deliver the most power when current and voltage limiting coincide. In transistor amplifiers this will be fixed, whereas valve amplifiers have output transformers and can deliver full power into a variety of load impedances simply by changing the transformer ratio.

2.8 Stereophony

As mentioned above, much greater spatial realism is obtained by using more than one loudspeaker. The most popular technique has been *stereophony*, nowadays abbreviated to stereo, based on two simultaneous audio channels feeding two spaced loudspeakers [2.8]. Figure 2.21 shows that the optimum listening arrangement for stereo is where the speakers and the listener are at different points of a triangle which is almost equilateral.

Stereophony works by creating differences in time of arrival of sound at the listener's ears. Section 1.6 showed that this is the most powerful hearing mechanism for determining direction. Figure 2.22(a) shows that this time of arrival difference is achieved by producing the same waveform at each speaker simultaneously, but with a difference in the relative level, rather than phase. Each ear picks up sound from both loudspeakers and sums the waveforms.

The sound picked up by the ear on the same side as the speaker is in advance of the same sound picked up by the opposite ear. When the level emitted by the left loudspeaker is greater than that emitted by the right, it will be seen from Figure 2.22(b) that the sum of the signals received at the left ear is a waveform which has arrived earlier than the sum of the waveforms received at the right ear. The time of arrival difference is interpreted as being due to a sound source left of centre.

The stereophonic illusion only works properly if the two loudspeakers are producing in-phase signals. In the case of an accidental phase reversal, the stereo will be ill-defined and give a spatial effect without images. At low frequencies the two loudspeakers are in each other's near field and so antiphase connection results in bass cancellation.

As the apparent position of a sound source between the two speakers can be controlled solely by the relative level of the sound emitted by each one, it is pos-

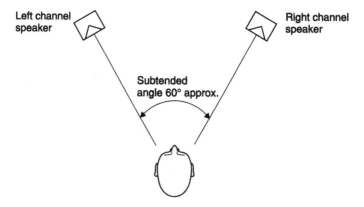

Figure 2.21 Configuration used for stereo listening.

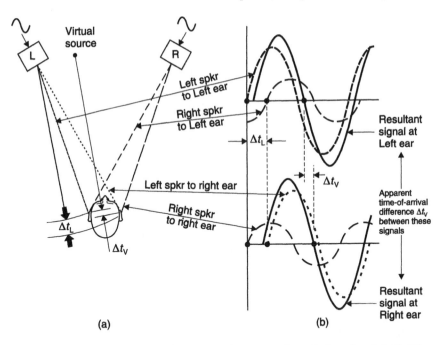

Figure 2.22(a) Stereo illusion is obtained by producing the same waveform at both speakers, but with different amplitudes. (b) As both ears hear both speakers but at different times, relative level causes apparent time shift at the listener. Δt_L = inter-aural delay due to loudspeaker; Δt_V = inter-aural delay due to virtual source.

sible to 'steer' a monophonic signal from a single microphone into a particular position in a stereo image using a form of differential gain control. Figure 2.23 shows that this device, known as a *panoramic potentiometer* or pan-pot for short, will produce equal outputs when the control is set to the centre. If the pan-pot is moved left or right, one output will increase and the other will reduce, moving or *panning* the stereo image to one side.

If the system is perfectly linear, more than one sound source can be panned into a stereo image, with each source heard in a different location. This is done using a stereo mixer, shown in Figure 2.24 in which monophonic inputs pass via pan-pots to a stereo mix bus. The majority of pop records are made in this way, usually with the help of a multi-track tape recorder with one track per microphone so that mixing and panning can be done at leisure.

2.9 Stereo microphones

Multi-track pan-potted audio can never be as realistic as the results of using a stereo microphone. The job of a stereo microphone is to produce two audio signals which have no phase differences but whose relative levels are a function of the direction from which sound arrives. The most spatially accurate technique involves the use of directional microphones which are coincidentally mounted but with their polar diagrams crossed. This configuration is known variously as a *crossed pair* or a *coincident pair*. Figure 2.25 shows a stereo microphone constructed by crossing a pair of figure-of-eight microphones at 90 degrees.

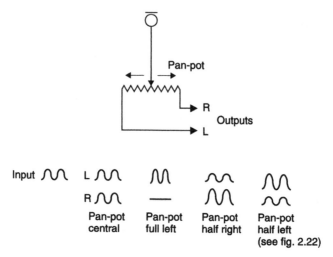

Figure 2.23 The pan-pot distributes a monophonic microphone signal into two stereo channels allowing the sound source to be positioned anywhere in the image.

The output from the two microphones will be equal for a sound source straight ahead, but as the source moves left, the output from the left-facing microphone will increase and the output from the right-facing microphone will reduce. When a sound source has moved 45 degrees off-axis, it will be in the response null of one of the microphones and so only one loudspeaker will emit sound. Thus the fully left or fully right reproduction condition is reached at ± 45 degrees. The angle between nulls in L and R is called the acceptance angle which has some parallels with the field of view of a camera.

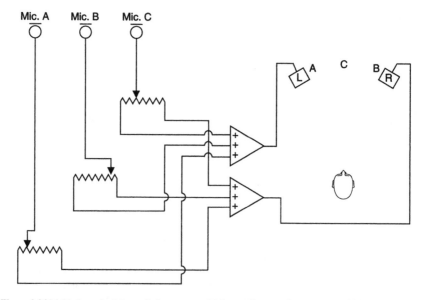

Figure 2.24 Multi-channel mixing technique pans multiple sound sources into one stereo image.

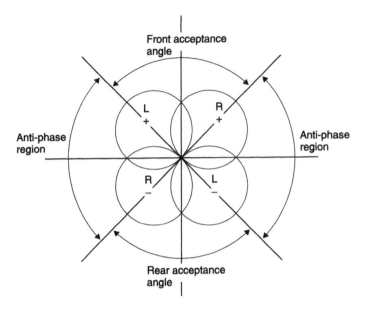

Figure 2.25 Crossed eight stereo microphone. Note acceptance angle between nulls.

Sounds between 45 and 135 degrees will be emitted out of phase and will not form an identifiable image. Important sound sources should not be placed in this region. Sounds between 135 and 225 degrees are in-phase and are mapped onto the frontal stereo image. The all-round pick-up of the crossed eight makes it particularly useful for classical music recording where it will capture the ambience of the hall.

Other polar diagrams can be used, for example the crossed cardioid, shown in Figure 2.26, is popular for applications where unwanted ambience needs to be suppressed. There is no obvious correct angle at which cardioids should be crossed, and the actual angle will depend on the application. Commercially available stereo microphones are generally built on the side-fire principle with one capsule vertically above the other. The two capsules can be independently rotated to any desired angle. Often the polar diagrams of the two capsules can be changed.

In the soundfield microphone, four capsules are fitted in a tetrahedron. By adding and subtracting proportions of these four signals in various ways it is possible to synthesize a stereo microphone having any acceptance angle and to point it in any direction relative to the body of the microphone. This can be done using the control box supplied with the microphone. Although complex the soundfield microphone has the advantage that it can be electrically steered and so no physical access is needed after it is slung. If all four outputs are recorded, the steering process can be performed in post production by connecting the control box to the recorder output.

Other techniques will be found in stereo, such as the use of spaced omnidirectional microphones. Although there is no scientific basis for the technique and precise images cannot be formed, a pleasing spacious sound can result.

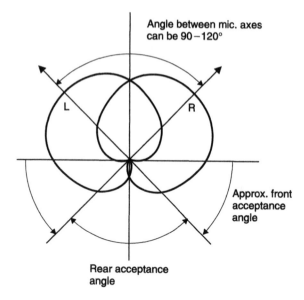

Figure 2.26 Crossed cardioid microphone.

Clearly the use of stereo will make it obvious if a microphone has been turned. In many applications in television sound the microphone is turned as a side effect of swinging a boom or fishpole, as in Figure 2.27 (a). This is undesirable in stereo, and new handling techniques are necessary to keep the microphone heading constant as in (b).

2.10 Headphone stereo

It should be clear that the result of Figure 2.22 cannot be obtained with headphones because these prevent both ears receiving both channels. As a result there is no stereophonic image and the sound appears, quite unrealistically, to be inside the listener's head. Consequently headphones are quite useless for monitoring the stereo image in signals designed for loudspeaker reproduction.

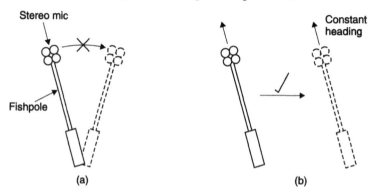

Figure 2.27 (a) Swinging a fishpole causes sound image to rotate. Tracking has to be used as in (b).

Highly realistic results can be obtained on headphones using the so-called *dummy head* microphone which is a more or less accurate replica of the human head with a microphone at each side. Clearly the two audio channels simply move the listener's ears to the location of the dummy. Unfortunately dummy head signals are incompatible with loudspeaker reproduction and are not used in television.

An improvement in headphone performance on signals intended for loudspeakers can be obtained using a shuffler [2.9]. This device, shown in Figure 2.28 simulates the cross-coupling of loudspeaker listening by feeding each channel to the other ear via a delay and a filter which simulates the effect of head shading. The result is a sound image which appears in front of the listener so that decisions regarding the spatial position of sources can be made. Although the advantages of the shuffler have been known for decades, the information appears to have eluded most equipment manufacturers.

2.11 M-S stereo

In television sound the apparent width of the stereo image may need to be adjusted, especially to obtain a good audio transition where there has been a change of shot or to match the sound stage to the picture. This can be done using *M-S stereo* and manipulating the difference between the two channels. Figure 2.29(a) shows that the two signals from the microphone, L and R, are passed through a *sum and difference* unit which produces two signals, M and S. The M or Mid signal is the sum of L and R, whereas the S or Side signal is the difference between L and R. The sums and differences are divided by two to keep the levels correct.

The result of this sum-difference process can be followed in Figure 2.29(b). A new polar diagram is drawn which represents the sum of L and R for all angles of approach. It will be seen that this results in a forward-facing eight, as if a monophonic microphone had been used, hence the term M or Mid. If the same process is performed using L–R, the result is a sideways-facing eight, hence the term S or Side. In L, R format the acceptance angle is between the nulls, whereas in M-S format the acceptance angle is between the points where the M and S polar diagrams cross.

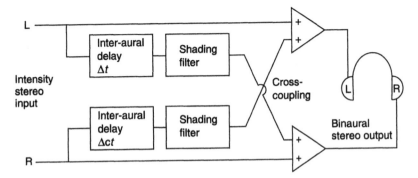

Figure 2.28 The shuffler converts L, R loudspeaker stereo into signals suitable for headphone reproduction by simulating the cross-coupling, delay and shading of loudspeaker reproduction.

The S signal can now be subject to variable gain. Following this a second sum and difference unit is used to return to L, R format for monitoring. The S gain control effectively changes the size of the S polar diagram without affecting the M polar diagram. Figure 2.29(c) shows that reducing the S gain makes the acceptance angle wider, whereas increasing the S gain makes it smaller. Clearly if the S gain control is set to unity, there will be no change to the signals.

Whilst M-S stereo can be obtained by using a conventional L, R microphone and a sum and difference network, it is clear from Figure 2.29(b) that M-S signals can be obtained directly using a suitable microphone. Figure 2.30 shows a number of M-S microphones in which the S capsule is always an eight. A variety of responses (other than omni) can be used for the M capsule.

The M-S microphone technique has a number of advantages. The narrowing polar diagram at high frequencies due to diffraction is less of a problem because the most prominent sound source will generally be in the centre of the stereo image and this is directly on the axis of the M capsule. An image width control can easily be built into an M-S microphone. A favourite mono microphone can be turned into an M-S microphone simply by mounting a side-facing eight above it.

On a practical note it is often necessary to use a stereo microphone with a fishpole. In some shots the microphone will be above the action, but in a close-up it may also be used below shot. Inverting the microphone in this way will interchange Left and Right channels. If an M-S microphone is used, the fishpole operator can operate a simple phase reverse switch in the S channel which will reverse the channels.

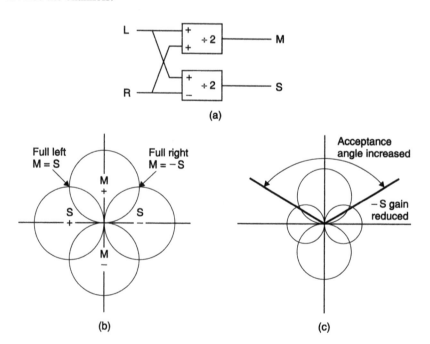

Figure 2.29 (a) M-S working adds and subtracts the L, R stereo signals to produce Mid and Side signals. In the case of a crossed eight, the M-S format is the equivalent of forward and sideways facing eights. (c) Changing the S gain alters the acceptance angle of the stereo microphone.

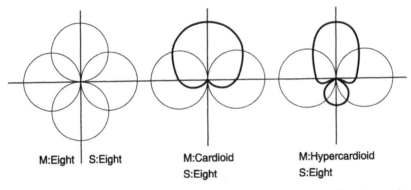

M:Eight | S:Eight   M:Cardioid   M:Hypercardioid
                      S:Eight   S:Eight

Figure 2.30 Various M-S microphone configurations, Note that the S microphone must always be an eight.

2.12 Mono compatibility

Whilst almost all consumer audio equipment is now stereo, the portable television set may well remain monophonic for some time to come and it will be necessary to consider the mono listener when making stereo material. There is a certain amount of compatibility between stereo and mono systems. If the S gain of a stereo signal is set to zero, only the M signal will pass. This is the component of the stereo image due to sounds from straight ahead and is the signal used when monophonic audio has to be produced from stereo. Sources positioned on the extreme edges of the sound stage will not appear as loud in mono as those in the centre and any antiphase ambience will cancel out, but in most cases the result is adequate.

Clearly an accidental situation in which one channel is phase reversed is catastrophic in mono as the centre of the image will be cancelled out.

One characteristic of stereo is that the viewer is able to concentrate on a sound coming from a particular direction using the cocktail party effect. Thus it will be possible to understand dialogue which is quite low in level even in the presence of other sounds in a stereo mix. In mono the viewer will not be able to use spatial discrimination and the result may be reduced intelligibility, which is particularly difficult for those with hearing impairments [2.10]. Consequently it is good practice to monitor stereo material in mono to check for acceptable dialogue.

A mono signal can also be reproduced on a stereo system by creating identical L and R signals, producing a central image only. Whilst there can be no real spatial information most people prefer mono on two speakers to mono on a single speaker.

2.13 Surround sound

Whilst the stereo illusion is very rewarding when well executed, it can only be enjoyed by a small audience and often a television set will be watched by more people than can fit into the acceptable stereo reproduction area. Surround sound is designed to improve the spatial representation of audio in film and television.

Figure 2.31 shows the 5.1 channel system proposed for advanced television sound applications. In addition to the conventional L and R stereo speakers at the

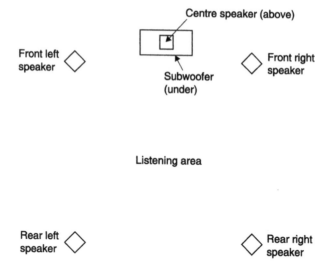

Figure 2.31 A 5.1 channel surround sound system.

front, a centre speaker is used. When normal L-R stereo is heard from off centre, the image will be pulled towards the nearer speaker. The centre speaker is primarily to pull central images back for the off-centre listener. In most television applications it is only the dialogue which needs to be centralized and consequently the centre speaker need not be capable of the full frequency range.

Rear L and R speakers are provided to allow reasonably accurate reproduction of sound sources from any direction, making a total of five channels. A narrow bandwidth sub-woofer channel is also provided to produce low frequencies for the inevitable earthquakes and explosions. The restricted bandwidth means that six full channels are not required, hence the term 5.1.

5.1 channel systems require the separate channels to be carried individually to the viewer. This is easy in digital compressed systems such as MPEG, but not in most consumer equipment such as VCRs which only have two audio channels. The Dolby Surround Sound system is designed to give some of the advantages of discrete multi-channel surround sound whilst only requiring two audio channels.

Figure 2.32(a) shows that Dolby Surround derives four channels, L, Centre, R and a surround channel which can be reproduced from two or more rear speakers. In fact a similar speaker arrangement to 5-channel discrete can be used. Figure 2.32(b) shows a Dolby Surround encoder. The centre channel is attenuated 3 dB and added equally to L and R. The surround signal is attenuated 3 dB and band limited prior to being encoded with a modified Dolby-B process. The resultant signal is then phase shifted so that a 180 degree phase relationship is formed between the components which are added to the Lt and Rt signals.

In a simple passive decoder, (c), Lt and Rt are used to drive L and R speakers directly. In the absence of a C speaker, the L and R speakers will reproduce the C signal centrally. The added antiphase surround signal will fail to produce an image. If a C speaker is provided, adding Lt and Rt will produce a suitable signal to drive it, although this will result in narrowing of the frontal image. Subtracting Lt and Rt will result in the sum of the surround signal and any difference

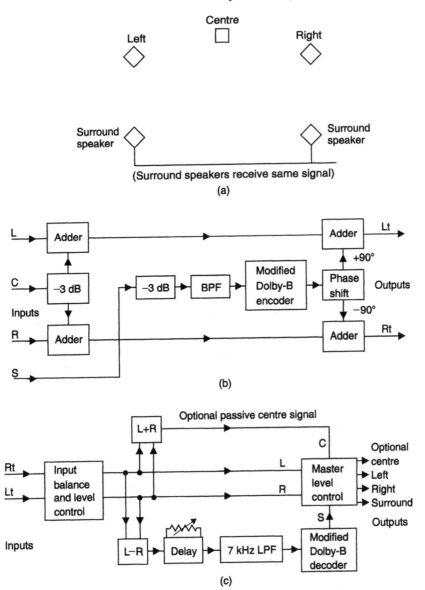

Figure 2.32 (a) Dolby surround sound speaker layout. (b) Dolby surround encoder; see text. (c) Simple passive decoder.

between the original L, R signals. This is band limited then passed through the Dolby-B decoder to produce a surround output.

The Dolby-B-like processing is designed to reduce audibility of L minus R breakthrough on the S signal, particularly on sibilants. The degree of compression is less than that in true Dolby-B to prevent difficulties when Lt and Rt are used as direct speaker drive signals.

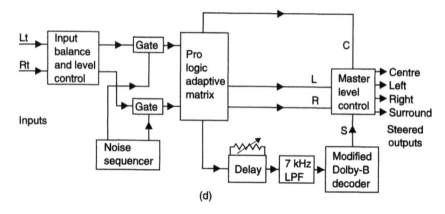

(d)

Figure 2.32(d) Pro-logic decoder uses variable matrices driven by analysis of dominant sound source location.

In the Pro-logic decoder, shown in Figure 2.32(d) the sum and difference stages are replaced with variable matrices which can act like a two-dimensional pan-pot or steering system. A simple decoder performs poorly when a single point source of sound is present in quiet surroundings, whereas a steered decoder will reduce the crosstalk output from unwanted channels. The steering is done by analysing the input signals to identify dominant sound sources. Figure 2.32(d) shows that comparison of the Lt and Rt signals will extract left/right dominance, whereas comparison of sum and difference signals will extract front/rear dominance.

2.14 Loudspeakers for stereo

The accuracy required for stereo reproduction is much greater than for mono. If there is any non-linearity in the system, different sound sources will intermodulate and produce phantom sound sources which appear to come from elsewhere in the stereo image than either of the original sounds. As these phantom sources are spatially separate from the genuine sources, they are easier to detect.

Where non-ideal speakers are used, it is important that the two speakers are absolutely identical. If the frequency and phase responses are not identical, the location of the apparent sound source will move with frequency. Where a harmonically rich source is involved, it will appear to be wider than it really is. This is known as *smear*.

If the loudspeakers suffer from beaming at HF then a proper stereo image will only be obtained at a small 'sweet spot' and all high frequencies will appear to emanate from the speakers, not from a central image. Small movements by the listener may cause quite large image changes. Irregular polar diagrams will also destroy stereo imaging. Such irregularities often occur near the crossover frequencies. Placing the drive units in a vertical line will prevent the horizontal polar diagram becoming too irregular. However, this idea is completely destroyed if such a speaker is placed on its side; a technique which is seen with depressing regularity. This may be because many loudspeakers are so mediocre that turning them on their side does not make them any worse.

Loudspeakers and amplifiers used for stereo mixing must have very low distortion and a precise tracking frequency and phase response. Loudspeakers must have smoothly changing directivity characteristics. In practice a great deal of equipment fails to meet these criteria.

In many applications in television the level of background noise is high and/or the room acoustic is deficient, making conventional monitoring difficult. In many cases there is little space left for loudspeakers once all of the video equipment is installed. One solution is the *close field* monitor which is designed and equalized so that the listener can approach very close to it. The term near field is often and erroneously used to describe close field monitors. The essence of close field monitoring is that direct sound reaches the listener so much earlier than the reverberant sound that the room acoustic becomes less important. In stereo close field monitoring the loudspeakers are much closer together, even though the same angle is subtended to the listener.

Figure 2.33 shows the APC-1000 loudspeaker (courtesy of Snell & Wilcox) which is a rack mounting stereo close field monitor especially designed for television applications. Whilst occupying only 4 rack units (7 inches or 18 cm), this unit uses active drive and rare-earth magnets to obtain response down to 30 Hz without stray magnetic fields. A line level crossover drives one amplifier per drive unit and these are aligned so that a correct stereo image will be obtained only 18 inches (45 cm away) away. In this case the direct sound also has significantly higher level than ambient sound so that monitoring can even take place in noisy surroundings.

Figure 2.33 In the APC-1000 loudspeaker, active technology is used to obtain full frequency range from only four rack units. Rare-earth magnets need no screening but do not disturb picture monitors.

References

2.1 Moir, J. Speaker directivity and sound quality. *Wireless World*, **85**, (1979)

2.2 Kelly, S. In Borwick, J (ed.) *Loudspeaker and Headphone Handbook*, Ch. 2. Oxford: Focal Press (1994)

2.3 Rice, C.W. and Kellog, E.W. Notes on the development of a new type of hornless loudspeaker. *J. Am. Inst. Elect. Engrs.*, **12**, 461–480 (1925)

2.4 Shorter, D.E.L. A survey of performance criteria and design considerations for high-quality monitoring loudspeakers. *Proc. Inst Elect Engrs*, **105**, Pt B, no. 24, 607–623 (1958)

2.5 Walker, P.J. New developments in electrostatic loudspeakers. *J. Audio Eng. Soc.*, **28** 795–799 (1980)

2.6 Olson, H.F. *Acoustical Engineering*, Philadelphia: Professional Audio Journals Inc. (1991)

2.7 Kates, J.M. Loudspeaker cabinet reflection effects. *J. Audio Eng. Soc.*, **27** (1979)

2.8 *Stereophonic Techniques*, AES Anthology. Audio Engineering Society (1986)

2.9 Thomas, M.V. Improving the stereo headphone sound image. *J. Audio Eng. Soc.*, **25**, 474-478 (1977)

2.10 Harvey, F.K. and Uecke, E.H., Compatibility problems in two channel stereophonic recordings. *J. Audio Eng. Soc.*, **10**, 8–12 (1962)

Chapter 3

Analog and digital audio signals

3.1 Audio measurements

The first audio signals to be transmitted were on telephone lines. Where the wiring is long compared to the electrical wavelength (not to be confused with the acoustic wavelength) of the signal, a transmission line exists in which the distributed series inductance and the parallel capacitance interact to give the line a characteristic impedance. In telephones this turned out to be about 600 Ω. In transmission lines the best power delivery occurs when the source and the load impedance are the same; this is the process of matching.

It was often required to measure the power in a telephone system, and 1 milliwatt was chosen as a suitable unit. Thus the reference against which signals could be compared was the dissipation of 1 mW in 600 Ωs. Figure 3.1(a) shows that according to Ohm's law the power dissipated in a resistance is proportional to the square of the applied voltage. This causes no difficulty with DC, but with alternating signals such as audio it is harder to calculate the power. Consequently a unit of voltage for alternating signals was devised. Figure 3.1(b) shows that the average power delivered during a cycle must be proportional to the mean of the square of the applied voltage. Since power is proportional to the square of applied voltage, the same power would be dissipated by a DC voltage whose value was equal to the square root of the mean of the square of the AC voltage. Thus the volt r.m.s. (root mean square) was specified. An AC signal of a given number of volts r.m.s. will dissipate exactly the same amount of power in a given resistor as the same number of volts DC.

Figure 3.2(a) shows that for a sine wave the r.m.s. voltage is obtained by dividing the peak voltage V pk by the square root of two. However, for a square wave (b) the r.m.s. voltage and the peak voltage are the same. Most moving coil AC voltmeters only read correctly on sine waves, whereas many electronic meters incorporate a true r.m.s. calculation.

On an oscilloscope it is often easier to measure the peak-to-peak voltage which is twice the peak voltage. The r.m.s. voltage cannot be measured directly on an oscilloscope since it depends on the waveform although the calculation is simple in the case of a sine wave.

Figure 3.1(a) also shows that the dissipation of 1 mW in 600 Ωs will be due to an applied voltage of 0.775 V r.m.s.. This voltage is the reference against which all audio levels are compared.

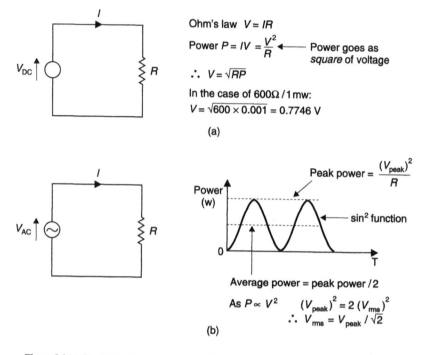

Figure 3.1 (a) Ohm's law: the power developed in a resistor is proportional to the square of the voltage. Consequently 1 mW in 600 Ω requires 0.775 V. With a sinusoidal alternating input (b), the power is a sine squared function which can be averaged over one cycle. A DC voltage which would deliver the same power has a value which is the square root of the mean of the square of the sinusoidal input.

3.2 The decibel

The decibel is a logarithmic measuring system and has its origins in telephony [3.1], where the loss in a cable is a logarithmic function of the length. Human hearing also has a logarithmic response with respect to sound pressure level (SPL). In order to relate to the subjective response, audio signal level measurements have also to be logarithmic and so the decibel was adopted for audio.

Figure 3.3 shows the principle of the logarithm. To give an example, if it is clear that 10^2 is 100 and 10^3 is 1000, then there must be a power between 2 and 3 to which 10 can be raised to give any value between 100 and 1000. That power is the logarithm to base 10 of the value. e.g. $\log_{10} 300 = 2.5$ approx. Note that 10^0 is 1.

Logarithms were developed by mathematicians before the availability of calculators or computers to ease calculations such as multiplication, squaring, division and extracting roots. The advantage is that armed with a set of log tables, multiplication can be performed by adding, division by subtracting. Figure 3.3 shows some examples. It will be clear that squaring a number is performed by adding two identical logs and the same result will be obtained by multiplying the log by 2.

The slide rule is an early calculator which consists of two logarithmically engraved scales in which the length along the scale is proportional to the log of the engraved number. By sliding the moving scale, two lengths can easily be

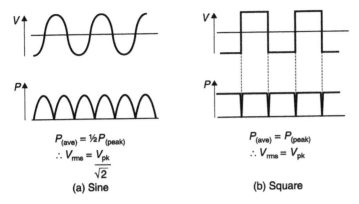

$$P_{(ave)} = \tfrac{1}{2}P_{(peak)}$$
$$\therefore V_{rms} = \frac{V_{pk}}{\sqrt{2}}$$
(a) Sine

$$P_{(ave)} = P_{(peak)}$$
$$\therefore V_{rms} = V_{pk}$$
(b) Square

Figure 3.2 (a) For a sine wave the conversion factor from peak to r.m.s. is $\sqrt{2}$. (b) For a square wave the peak and r.m.s. voltage is the same.

added or subtracted and as a result multiplication and division is readily obtained.

The logarithmic unit of measurement in telephones was called the B after Alexander Graham Bell, the inventor. Figure 3.4(a) shows that the B was defined as the log of the *power* ratio between the power to be measured and some

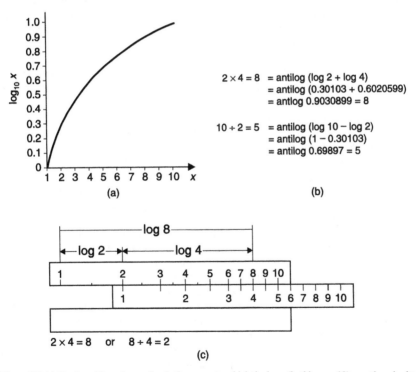

$2 \times 4 = 8$ = antilog (log 2 + log 4)
= antilog (0.30103 + 0.6020599)
= antilog 0.9030899 = 8

$10 \div 2 = 5$ = antilog (log 10 − log 2)
= antilog (1 − 0.30103)
= antilog 0.69897 = 5

(a)   (b)

$2 \times 4 = 8$ or $8 \div 4 = 2$

(c)

Figure 3.3 (a) The logarithm of a number is the power to which the base (in this case 10) must be raised to obtain the number. (b) Multiplication is obtained by adding logs, division by subtracting. (c) The slide rule has two logarithmic scales whose length can easily be added or subtracted.

$$1 \text{ bel} = \log_{10} \frac{P_1}{P_2} \quad 1 \text{ decibel} = \text{\small 1/10 bel}$$

$$\text{Power ratio (dB)} = 10 \times \log_{10} \frac{P_1}{P_2}$$

(a)

As power $\propto V^2$, when using voltages:

$$\text{Power ratio (dB)} = 10 \times \log \frac{V_1^2}{V_2^2}$$

$$= 10 \times \log \frac{V_1}{V_2} \times 2$$

$$= 20 \log \frac{V_1}{V_2}$$

(b)

Figure 3.4 (a) The bel is the log of the ratio between two powers, that to be measured and the reference. The bel is too large so the decibel is used in practice. (b) As the dB is defined as a power ratio, voltage ratios have to be squared. This is conveniently done by doubling the logs so the ratio is now multiplied by 20.

reference power. Clearly the reference power must have a level of 0 bels since $\log_{10} 1$ is 0.

The bel was found to be an excessively large unit for many purposes and so it was divided into 10 decibels, abbreviated dB, with a small d and a large B and pronounced deebee. Consequently the number of dB is 10 times the log of the power ratio. A device such as an amplifier can have a fixed power gain which is independent of signal level and this can be measured in dB. However, when measuring the power of a signal, it must be appreciated that the dB is a ratio and to quote the number of dBs without stating the reference is about as senseless as describing the height of a mountain as 2000 without specifying whether this is feet or metres. To show that the reference is 1 mW into 600 Ωs, the units will be dB(m). In radio engineering, the dB(W) will be found which is power relative to 1 watt.

Although the dB(m) is defined as a power ratio, level measurements in audio are often done by measuring the signal voltage using 0.775 V as a reference in a circuit whose impedance is not necessarily 600 Ω. Figure 3.3(b) shows that as the power is proportional to the square of the voltage, the power ratio will be obtained by squaring the voltage ratio. As squaring in logs is performed by doubling, the squared term of the voltages can be replaced by multiplying the log by a factor of two. To give a result in decibels, the log of the voltage ratio now has to be multiplied by 20.

Whilst 600 Ω matched impedance working is essential for analog telephones, it is quite inappropriate for audio wiring in a studio. The wavelength of audio in wires at 20 kHz is 15 kilometres. Most studios are built on a smaller scale than this and impedance matching is undesirable. Figure 3.5(a) shows that when impedance matching is required the output impedance of a signal source must be raised so that a potential divider is formed with the load. The actual drive voltage must be twice that needed on the cable as the potential divider effect wastes 6 dB of signal level and requires unnecessarily high power supply rail voltages in equipment. A further problem is that cable capacitance can cause an undesirable HF roll-off in conjunction with the high source impedance.

In modern professional audio equipment, shown in Figure 3.5(b), the source has the lowest output impedance practicable. This means that any ambient inter-

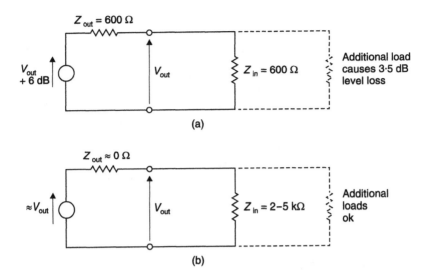

Figure 3.5 (a) Traditional impedance matched source wastes half the signal voltage in the potential divider due to the source impedance and the cable. (b) Modern practice is to use low output impedance sources with high impedance loads.

ference is attempting to drive what amounts to a short circuit and can only develop very small voltages. Furthermore shunt capacitance in the cable has very little effect. The destination has a somewhat higher impedance (generally a few kilohms) to avoid excessive currents flowing and to allow several loads to be placed across one driver.

In the absence of a fixed impedance it is now meaningless to consider power. Consequently only signal voltages are measured. The reference remains at 0.775 V, but power and impedance are irrelevant. Voltages measured in this way are expressed in dB(u); the commonest unit of level in modern systems. Most installations boost the signals on interface cables by 4 dB. As the gain of receiving devices is reduced by 4 dB, the result is a useful noise advantage without risking distortion due to the drivers having to produce high voltages.

In order to make the difference between dB(m) and dB(u) clear, consider the lossless matching transformer shown in Figure 3.6. The turns ratio is 2:1, therefore the impedance matching ratio is 4:1. As there is no loss in the transformer, the power in is the same as the power out so that the transformer shows a gain of 0 dB(m). However, the turns ratio of 2:1 provides a voltage gain of 6 dB(u). The doubled output voltage will develop the same power into the quadrupled load impedance.

In a complex system signals may pass through a large number of processes, each of which may have a different gain. Figure 3.7 shows that if one stays in the linear domain and measures the input level in volts r.m.s., the output level will be obtained by multiplying by the gains of all of the stages involved. This is a complex calculation.

The difference between the signal level with and without the presence of a device in a chain is called the *insertion loss,* measured in dB. However, if the input is measured in dB(u), the output level of the first stage can be obtained by adding the insertion loss in dB. The output level of the second stage can be obtained by further adding the loss of the second stage in dB and so on. The final result is obtained by adding together all of the insertion losses in dB and adding them to the input level in dB(u) to give the output level in dB(u). As the dB is a pure ratio it can multiply anything (by addition of logs) without changing the units. Thus dB(u) of level added to dB of gain are still dB(u).

In acoustic measurements, the sound pressure level (SPL) is measured in decibels relative to a reference pressure of 2×10^5 Pa r.m.s.. In order to make the reference clear the units are dB(SPL). In measurements which are intended to convey an impression of subjective loudness, a weighting filter is used prior to the level measurement which reproduces the frequency response of human hearing which is most sensitive in the mid-range. The most common standard frequency response is the so-called A-weighting filter, hence the term dB(A) used when a weighted level is being measured. At high or low frequencies, a lower reading will be obtained in dB(A) than in dB(SPL).

3.3 Audio level metering

There are two main reasons for having level meters in audio equipment: to line up or adjust the gain of equipment, and to assess the amplitude of the programme material.

Line up is often done using a 1 KHz sine wave generated at an agreed level such as 0 dB(u). If a receiving device does not display the same level, then its input sensitivity must be adjusted. Tape recorders and other devices which pass

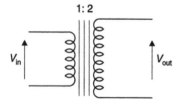

$P_{out} = P_{in}$ ∴ gain dB(m) = 0 dB(m)
$V_{out} = 2 \times V_{in}$ ∴ gain dB(u) = 6 dB(u)
$Z_{out} = 4 \times Z_{in}$

Figure 3.6 A lossless transformer has no power gain so the level in dB(m) on input and output is the same. However there is a voltage gain when measurements are made in dB(u).

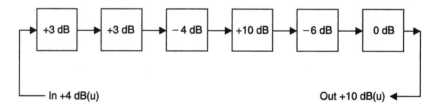

In +4 dB(u)   Out +10 dB(u)

Figure 3.7 In complex systems each stage may have voltage gain measured in dB. By adding all of these gains together and adding to the input level in dB(u), the output level in dB(u) can be obtained.

signals through are usually lined up so that their input and output levels are identical, i.e. their insertion loss is 0 dB. Line up is important in large systems because it ensures that inadvertent level changes do not occur.

In measuring the level of a sine wave for the purposes of line up, the dynamics of the meter are of no consequence, whereas on programme material the dynamics matter a great deal. The simplest (and cheapest) level meter is essentially an AC voltmeter with a logarithmic response. As the ear is logarithmic, the deflection of the meter is roughly proportional to the perceived volume, hence the term volume unit (VU) meter.

In audio recording and broadcasting, the worst sin is to overmodulate the tape or the transmitter by allowing a signal of excessive amplitude to pass. Real audio signals are rich in short transients which pass before the sluggish VU meter responds. Consequently the VU meter is also called the virtually useless meter in professional circles.

Broadcasters developed the peak program meter (PPM) which is also logarithmic, but which is designed to respond to peaks as quickly as the ear responds to distortion. Consequently the attack time of the PPM is carefully specified. If a peak is so short that the PPM fails to indicate its true level, the resulting overload will also be so brief that the ear will not hear it. A further feature of the PPM is that the decay time of the meter is very slow, so that any peaks are visible for much longer and the meter is easier to read because the meter movement is less violent.

The original PPM as developed by the BBC was sparsely calibrated, but other users have adopted the same dynamics and added decibel scales. Figure 3.8 shows some of the scales in use.

In broadcasting, the use of level metering and line up procedures ensures that the level experienced by the viewer does not change significantly from programme to programme. Consequently in a transmission suite, the goal would be to broadcast tapes at a level identical to that which was obtained during production. However, when making a recording prior to any production process, the goal would be to modulate the tape as fully as possible without clipping as this would then give the best signal-to-noise ratio. The level would then be reduced if necessary in the production process.

3.4 Stereo metering

In stereo systems is it important that the left and right channels display the same gain after line up. It is also important that the left and right channels are not inadvertently exchanged, and that that both channels have the same polarity. Often an indication of the width of the stereo image is useful. In some stereo equipment a *twin PPM* is fitted, having two needles which operate coaxially. One is painted red (L) and the other green (R). In stereo line up tone, the left channel is often interrupted briefly so that it can be distinguished from the right channel. The interruptions are so brief that the PPM reading is unaffected.

Unfortunately the twin PPM gives no indication that the unacceptable out-of-phase condition exists. A better solution is the *twin-twin PPM* which is two coaxial PPMs, one showing L, R and one showing M, S (see Section 2.11). When lining up for identical channel gain, obtaining an S null is more accurate. Some meters incorporate an S gain boost switch so that a deeper null can be displayed.

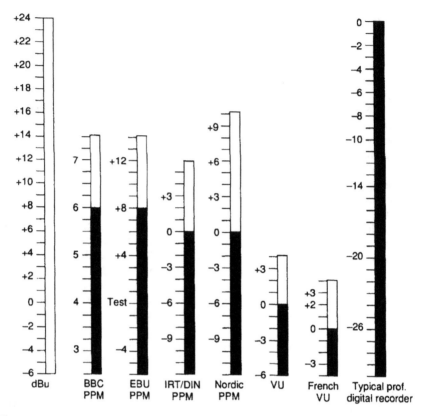

Figure 3.8 Some of the scales used in conjunction with the PPM dynamics. (After David Pope, with permission.)

When there is little stereo width, the M reading will exceed the S reading. Equal M and S readings indicate a strong source at one side of the sound stage. When an antiphase condition is met, the S level will exceed the M level. The M needle is usually white, and the S needle is yellow. This is not very helpful under dim incandescent lighting which makes both appear yellow. Exasperated users sometimes lever the front off the meter and put black stripes on the S needle. In modern equipment the moving coil meter is giving way to the bargraph meter which is easier to read.

The audio vectorscope is a useful tool which gives a lot of spatial information although it is less useful for level measurements. If an oscilloscope is connected in X, Y mode so that the M signal causes vertical beam deflection and the S signal causes lateral deflection, Figure 3.9 shows that the result will be a trace which literally points to the dominant sound sources in the stereo image. Visual estimation of the width of the stereo image is possible. An out-of-phase condition causes the trace to become horizontal. Non-imaging stereo from, for example, spaced microphones causes the trace to miss the origin because of phase differences between the channels.

Whilst self-contained audio vectorscopes are available, in television it is possible to employ a unit which synthesizes a video signal containing the

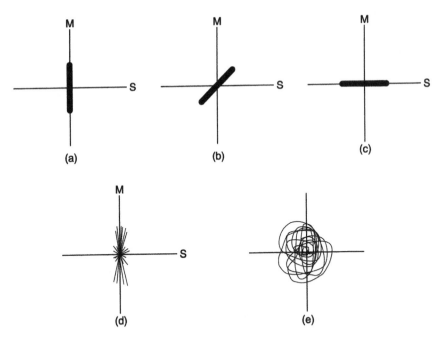

Figure 3.9 An audio vectorscope deflects the beam vertically for M inputs and horizontally for S inputs. (a) Mono signal L = R. (b) Fully right. (c) Antiphase condition. (d) Coincident microphones with dominant central image. (e) Spaced omni microphones.

vectorscope picture. This can then be keyed into the video signal to a convenient picture monitor. Some units also provide L, R, M, S bargraphs or virtual meters in the video.

3.5 Analog audio interfacing

Balanced line working was developed for professional audio as a means to reject noise. Figure 3.10 shows how balanced audio should be connected. The receiver subtracts one input from the other which rejects any common mode noise or hum picked up on the wiring. Twisting the wires tightly together ensures that both pick up the same amount of interference.

The standard connector which has been used for professional audio for many years is the XLR which has three pins. It is easy to remember that pins 1,2 and 3 connect to eXternal, Live and Return respectively. External is the cable screen, Live is the in-phase leg of the balanced signal and return is self explanatory. The metal body shell of the XLR connector should be connected to both the cable screen and pin 1 although cheaper connectors do not provide a tag for the user to make this connection and rely on contact with the chassis socket to ground the shell. Oddly, the male connector (the one with pins) is used for equipment signal outputs, whereas the female (the one with receptacles) is used with signal inputs. This is so that when phantom power is used, the live parts are insulated.

When making balanced cables it is important to ensure that the twisted pair is connected identically at both ends. If the two wires are inadvertently

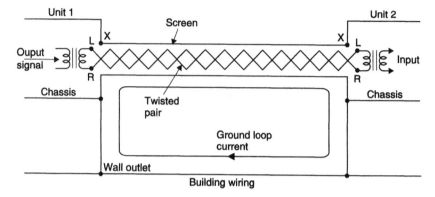

Figure 3.10 Balanced analog audio interface. Note that the braid plays no part in transmitting the audio signal, but bonds the two units together and acts as a screen. Loop currents flowing in the screen are harmless.

interchanged, the cable will still work, but a phase reversal will result, causing problems in stereo installations.

The screen does not carry the audio, but serves to extend the screened cabinets of the two pieces of equipment with what is effectively a metallic tunnel. For this to be effective against RF interference it has to be connected at both ends. This is also essential for electrical safety so that no dangerous potential difference can build up between the units. Figure 3.10 also shows that connecting the screen at both ends causes an earth loop with the building ground wiring. Loop currents will circulate as shown. This is not a problem because by shunting loop currents into the screen, they are kept out of the audio wiring.

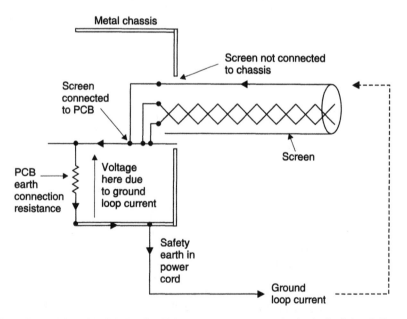

Figure 3.11 Poorly designed product in which screen currents pass to chassis via circuit board. Currents flowing in ground lead will raise voltages which interfere with the audio signal.

Some poorly designed equipment routes the X-pin of the XLR via the PCB instead of direct to the equipment frame. As Figure 3.11 shows, this effectively routes loop currents through the circuitry and is prone to interference. This approach will not pass recent EMC regulations but there is a lot of old equipment still in service which should be put right. A simple track cut and a new chassis bonded XLR socket is often all that is necessary. Another false economy is the use of plastic XLR shells which cannot provide continuity of screening.

Differential working with twisted pairs is designed to reject hum and noise, but it only works properly if both signal legs have identical frequency/impedance characteristics at both ends. The easiest way of achieving this is to use transformers which give much better RF rejection than electronic balancing. Whilst an effective electronic differential receiver can be designed with care, a floating balanced electronic driver cannot compete with a transformer.

In consumer equipment, differential working is considered too expensive. With single ended analog signals using coaxial (coax) cable as found on 'phono', DIN and single pole jack connectors, effective transmission over long distances is very difficult. When the signal return, the chassis ground and the safety ground are one and the same as in Figure 3.12(a), ground loop currents cannot be rejected. The only solution is to use equipment which is double insulated so that no safety ground is needed. Then each item can be grounded by the coax screen. As Figure 3.12(b) shows, there can then be no ground current as there is no loop. However, unbalanced working also uses higher impedances and lower signal levels and is more prone to interference.

3.6 Phantom power

Where active microphones (those containing powered circuitry) are used, it is common to provide the power using the balanced audio cable in reverse. Figure 3.13 shows how phantom powering works. As the audio signal is transmitted differentially, the DC power can be fed to the microphone without interfering with the returning audio signal. The use of female connectors for audio inputs is because of phantom power. An audio input is a phantom power output and so requires the insulated contacts of the female connector. Provision is generally made to turn the phantom power off so that dynamic passive microphones can be used.

3.7 Introduction to digital audio

In an analog system, information is conveyed by some infinite variation of a continuous parameter such as the voltage on a wire or the strength of flux on a tape. In a recorder, distance along the medium is a further, continuous, analog of time.

Those characteristics are the main weakness of analog signals. Within the allowable bandwidth, *any* waveform is valid. If the speed of the medium is not constant, one valid waveform is changed into another valid waveform; a timebase error cannot be detected in an analog system. In addition, a voltage error simply changes one valid voltage into another; noise cannot be detected in an analog system. It is a characteristic of analog systems that degradations cannot be

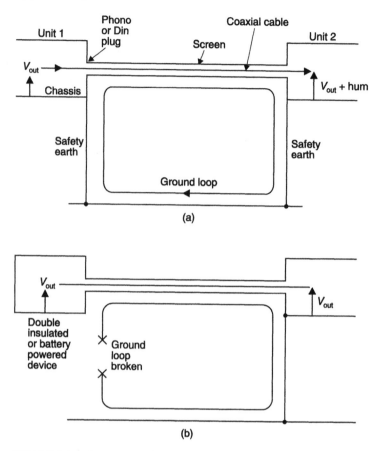

Figure 3.12 (a) Unbalanced consumer equipment cannot be protected from hum loops because the signal return and the screen are the same conductor. (b) With a floating signal source there will be no current in the screen. Source must be double insulated for safety.

separated from the original signal, so nothing can be done about them. At the end of a system a signal carries the sum of all degradations introduced at each stage through which it passed. This sets a limit to the number of stages through which a signal can be passed before it is useless.

An ideal digital audio channel has the same characteristics as an ideal analog channel: both of them are totally transparent and reproduce the original applied waveform without error. Needless to say, in the real world ideal conditions seldom prevail, so analog and digital equipment both fall short of the ideal. Digital audio simply falls short of the ideal by a smaller distance than does analog and at lower cost, or, if the designer chooses, can have the same performance as analog at much lower cost.

There is one system, known as pulse code modulation (PCM), which is in virtually universal use. Figure 3.14 shows how PCM works. The time axis is represented in a discrete or stepwise manner and the waveform is carried by measurement at regular intervals. This process is called sampling and the frequency with which samples are taken is called the sampling rate or sampling frequency F_s. The sampling rate is generally fixed and every effort is made to rid the sampling clock of jitter so that every sample will be made at an exactly even time step. If there is any subsequent timebase error, the instants at which samples arrive will be changed, but the effect can be eliminated by storing the samples temporarily in a memory and reading them out using a stable, locally generated clock. This process is called timebase correction and all properly engineered digital audio systems must use it. Clearly timebase error is not reduced; it is totally eliminated. As a result there is little point measuring the wow and flutter of a digital recorder; it does not have any.

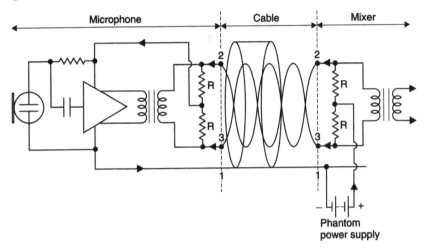

Figure 3.13 Phantom power system allows mixing console to power microphone down signal cable.

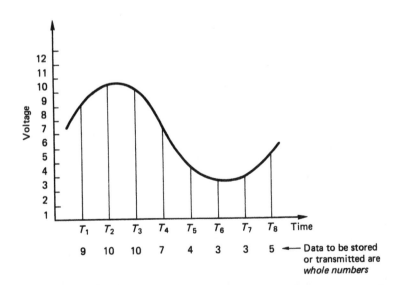

Figure 3.14 In pulse code modulation the analog waveform is measured periodically at the sampling rate. The voltage (represented here by the height) of each sample is then described by a whole number. The whole numbers are stored or transmitted rather than the waveform itself.

Figure 3.14 also shows that each sample is also discrete, or represented in a stepwise manner. The length of the sample, which will be proportional to the voltage of the audio waveform, is represented by a whole number. This process is known as quantizing and results in an approximation, but the size of the error can be controlled by using more steps until it is negligible. The advantage of using whole numbers is that they are not prone to drift. If a whole number can be carried from one place to another without numerical error, it has not changed at all. By describing audio waveforms numerically, the original information has been expressed in a way which is better able to resist unwanted changes.

Essentially, digital audio carries the original waveform numerically. The number of the sample is an analog of time, and the magnitude of the sample is an analog of the pressure at the microphone.

As both axes of the digitally represented waveform are discrete, the waveform can be accurately restored from numbers as if it were being drawn on graph paper. If we require greater accuracy, we simply choose paper with smaller squares. Clearly more numbers are then required and each one could change over a larger range.

Digital audio has two main advantages, but it is not possible to say which is the most important:

a) The quality of reproduction of a well-engineered digital audio system is independent of the medium and depends only on the quality of the conversion processes and any compression technique.

b) The conversion of audio to the digital domain allows tremendous opportunities which were denied to analog signals.

Someone who is only interested in sound quality will judge the former to be the most relevant. If good quality convertors can be obtained, all of the shortcomings of analog recording can be eliminated to great advantage. When a digital recording is copied, the same numbers appear on the copy: it is not a dub, it is a clone. If the copy is indistinguishable from the original, there has been no generation loss. Digital recordings can be copied indefinitely without loss of quality.

Once audio is in the digital domain, it becomes data, and as such is indistinguishable from any other type of data. Systems and techniques developed in other industries for other purposes can be used for audio. Computer equipment is available at low cost because the volume of production is far greater than that of professional audio equipment. Disk drives and memories developed for computers can be put to use in audio products. A word processor adapted to handle audio samples becomes a workstation. There seems to be little point in waiting for a tape to wind when a disk head can access data in milliseconds. The difficulty of locating the edit point and the irrevocable nature of tape-cut editing are hardly worth considering when the edit point can be located by viewing the audio waveform on a screen or by listening at any speed to audio from a memory. The edit can be simulated and trimmed before it is made permanent.

Communications networks developed to handle data can happily carry digital audio over indefinite distances without quality loss. Digital audio broadcasting (DAB) makes use of these techniques to eliminate the interference, fading and multipath reception problems of analog broadcasting. At the same time, more efficient use is made of available bandwidth.

3.8 Binary

Binary is the most minimal numbering system, which has only two digits, 0 and 1. BInary digiTS are universally contracted to bits. These are readily conveyed in switching circuits by an 'on' state and an 'off' state. With only two states, there is little chance of error.

Figure 3.15 shows that in binary, the bits represent one, two, four, eight, sixteen etc. A multi-digit binary number is commonly called a word, and the number of bits in the word is called the wordlength. The right-hand bit is called the least significant bit (LSB), whereas the bit on the left-hand end of the word is called the most significant bit (MSB). Clearly more digits are required in binary than in decimal, but they are more easily handled. A word of eight bits is called a byte, which is a contraction of 'by eight'.

In a digital audio system, the length of the sample is expressed by a binary number of typically 16 bits. The signals sent have two states, and change at predetermined times according to some stable clock. Figure 3.16 shows the consequences of this form of transmission. If the binary signal is degraded by noise, this will be rejected by the receiver, which judges the signal solely by whether it is above or below the halfway threshold, a process known as slicing. The signal will be carried in a channel with finite bandwidth, and this limits the slew rate of the signal; an ideally upright edge is made to slope. Noise added to a sloping signal can change the time at which the slicer judges that the level passed through the threshold. This effect is also eliminated when the output of the slicer is reclocked. However many stages the binary signal passes through, it still comes out the same, only later.

Audio samples which are represented by whole numbers can be carried reliably from one place to another by such a scheme, and if the number is correctly received, there has been no loss of information en route.

There are two ways in which binary signals can be used to carry audio samples and these are shown in Figure 3.17. When each digit of the binary number is carried on a separate wire this is called parallel transmission. The state of the

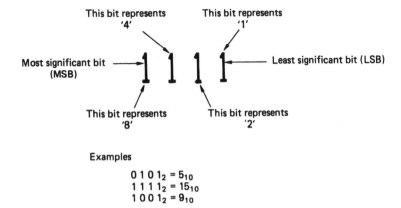

Examples

$$0\ 1\ 0\ 1_2 = 5_{10}$$
$$1\ 1\ 1\ 1_2 = 15_{10}$$
$$1\ 0\ 0\ 1_2 = 9_{10}$$

Figure 3.15 In a binary number, the digits represent increasing powers of two from the LSB. Also defined here are MSB and wordlength. When the wordlength is 8 bits, the word is a byte. Binary numbers are used as memory addresses, and the range is defined by the address wordlength. Some examples are shown here.

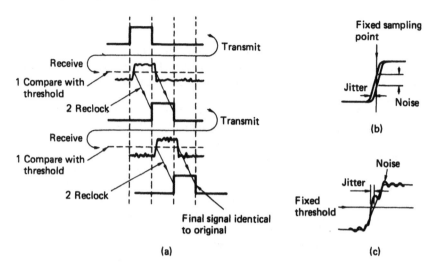

Figure 3.16 (a) A binary signal is compared with a threshold and reclocked on receipt, thus the meaning will be unchanged. (b) Jitter on a signal can appear as noise with respect to fixed timing. (c) Noise on a signal can appear as jitter when compared with a fixed threshold.

wires changes at the sampling rate. Using multiple wires is cumbersome, particularly where a long wordlength is in use, and a single wire can be used where successive digits from each sample are sent serially. This is the definition of PCM. Clearly the clock frequency must now be higher than the sampling rate. Whilst digital transmission of audio eliminates noise and timebase error, there is a penalty that a single high quality audio channel requires around 1 million bits per second. Clearly digital audio could only come into use when such a data rate could be handled economically. Further applications become possible when means to reduce the data rate become economic.

3.9 Conversion

The input to a converter is a continuous-time, continuous-voltage waveform, and this is changed into a discrete-time, discrete-voltage format by a combination of sampling and quantizing. These two processes are totally independent and can be

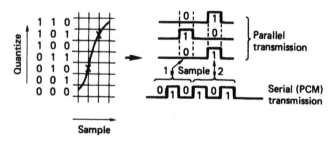

Figure 3.17 When a signal is carried in numerical form, either parallel or serial, the mechanisms of Figure 3.16 ensure that the only degradation is in the conversion processes.

performed in either order and discussed quite separately in some detail. Figure 3.18(a) shows an analog sampler preceding a quantizer, whereas (b) shows an asynchronous quantizer preceding a digital sampler. Ideally, both will give the same results; in practice each has different advantages and suffers from different deficiencies. Both approaches will be found in real equipment.

3.10 Sampling and aliasing

The sampling process originates with a regular pulse train which is shown in Figure 3.19(a) to be of constant amplitude and period. The audio waveform amplitude-modulates the pulse train in much the same way as the carrier is modulated in an AM radio transmitter. One must be careful to avoid over-modulating

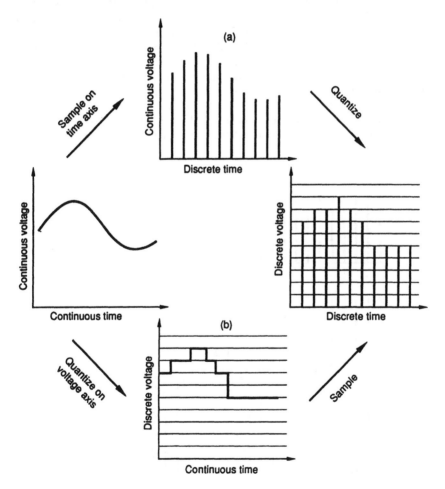

Figure 3.18 Since sampling and quantizing are orthogonal, the order in which they are performed is not important. In (a) sampling is performed first and the samples are quantized. This is common in audio converters. In (b) the analog input is quantized into an asynchronous binary code. Sampling takes place when this code is latched on sampling clock edges. This approach is universal in video converters.

the pulse train as shown in (b) and this is achieved by applying a DC offset to the analog waveform so that silence corresponds to a level halfway up the pulses, as at (c). Clipping due to any excessive input level will then be symmetrical.

In the same way that AM radio produces sidebands or images above and below the carrier, sampling also produces sidebands although the carrier is now a pulse train and has an infinite series of harmonics as shown in Figure 3.20(a). The sidebands repeat above and below each harmonic of the sampling rate as shown in (b).

The sampled signal can be returned to the continuous-time domain simply by passing it into a low-pass filter. This filter has a frequency response which prevents the images from passing, and only the baseband signal emerges, completely unchanged. If considered in the frequency domain, this filter is called an anti-image or reconstruction filter.

If an input is supplied having an excessive bandwidth for the sampling rate in use, the sidebands will overlap, (Figure 3.20(c)) and the result is aliasing, where certain output frequencies are not the same as their input frequencies but instead become difference frequencies (d). It will be seen from Figure 3.20 that aliasing does not occur when the input frequency is equal to or less than half the sampling rate, and this derives the most fundamental rule of sampling, which is that the sampling rate must be at least twice the highest input frequency.

Whilst aliasing has been described above in the frequency domain, it can be described equally well in the time domain. In Figure 3.21(a) the sampling rate is obviously adequate to describe the waveform, but at (b) it is inadequate and aliasing has occurred.

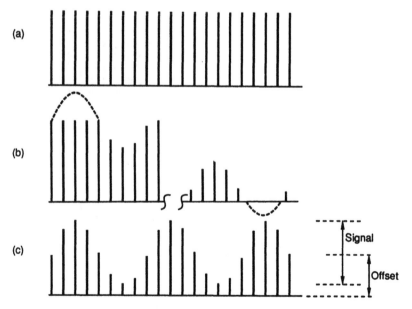

Figure 3.19 The sampling process requires a constant-amplitude pulse train as shown in (a). This is amplitude modulated by the waveform to be sampled. If the input waveform has excessive amplitude or incorrect level, the pulse train clips as shown in (b). For an audio waveform, the greatest signal level is possible when an offset of half the pulse amplitude is used to centre the waveform as shown in (c).

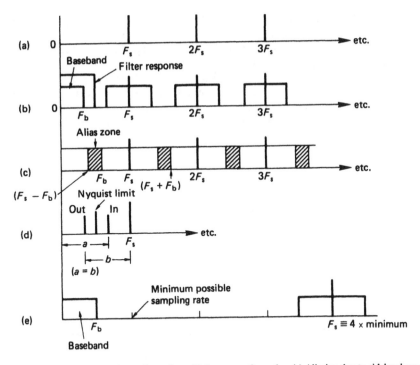

Figure 3.20 (a) Spectrum of sampling pulses. (b) Spectrum of samples. (c) Aliasing due to sideband overlap. (d) Beat-frequency production. (e) 4 × oversampling.

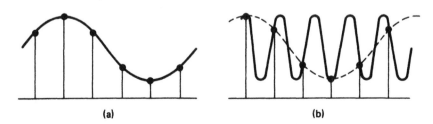

Figure 3.21 In (a) the sampling is adequate to reconstruct the original signal. In (b) the sampling rate is inadequate, and reconstruction produces the wrong waveform (dashed). Aliasing has taken place.

In practice it is necessary also to have a low-pass, or anti-aliasing filter at the input to prevent frequencies of more than half the sampling rate from reaching the sampling stage.

If ideal low-pass anti-aliasing and anti-image filters are assumed, an ideal spectrum shown at Figure 3.22(a) is obtained. The impulse response of a phase-linear ideal low-pass filter is a sin x/x waveform in the time domain, and this is shown in Figure 3.22(b). Such a waveform passes through zero volts periodically. If the cut-off frequency of the filter is one-half of the sampling rate, the impulse passes through zero *at the sites of all other samples*. It can be seen from Figure 3.22(c) that at the output of such a filter the voltage at the centre of a sample is due to that sample alone, since the value of *all* other samples is zero at that instant. In other words the continuous time output waveform must pass through the peaks of the input samples. In between the sample instants, the output of the filter is the sum of the contributions from many impulses, and the waveform smoothly joins the tops of the samples.

The ideal filter with a vertical 'brick-wall' cut-off slope is impossible to implement and in practice a filter with a finite slope has to be accepted, as shown in Figure 3.23. The cut-off slope begins at the edge of the required band, and consequently the sampling rate has to be raised a little to drive aliasing products to an acceptably low level.

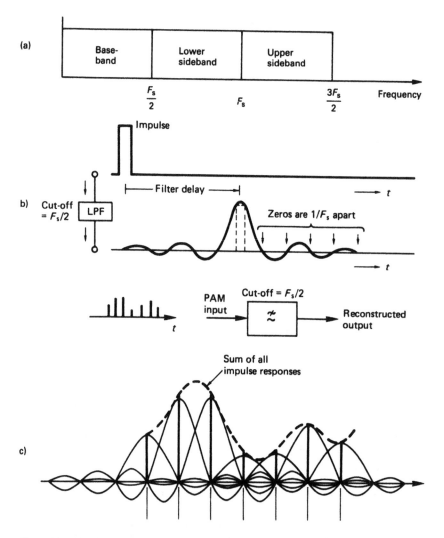

Figure 3.22 If ideal 'brick wall' filters are assumed, the efficient spectrum of (a) results. An ideal low pass filter has an impulse response shown in (b). The impulse passes through zero at intervals equal to the sampling period. When convolved with a pulse train at the sampling rate, as shown in (c), the voltage at each sample instant is due to that sample alone as the impulses from all other samples pass through zero there.

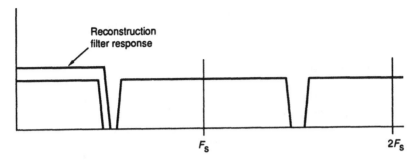

Figure 3.23 As filters with finite slope are needed in practical systems, the sampling rate is raised slightly beyond twice the highest frequency in the baseband.

Every signal which has been through the digital domain has passed through both an anti-aliasing filter and a reconstruction filter. These filters must be carefully designed in order to prevent audible artifacts, particularly those due to lack of phase linearity, as they may be audible [3.2,3,4]. The nature of the filters used has a great bearing on the subjective quality of the system.

Much effort can be saved in analog filter design by using oversampling which is considered later in this chapter. Strictly oversampling means no more than that a higher sampling rate is used than is required by sampling theory. In the loose sense an 'oversampling converter' generally implies that some combination of high sampling rate and various other techniques has been applied. The audible superiority and economy of oversampling converters has led them to be almost universal.

3.11 Choice of sampling rate

At one time, video recorders were adapted to store audio samples by creating a pseudo-video waveform which could convey binary as black and white levels. The sampling rate of such a system is constrained to relate simply to the field rate and field structure of the television standard used, so that an integer number of samples can be stored on each usable television line in the field. In 60 Hz video, there are 35 blanked lines, leaving 490 lines per frame, or 245 lines per field for samples. If three samples are stored per line, the sampling rate becomes $60 \times 245 \times 3 = 44.1$ kHz. This sampling rate was adopted for Compact Disc. Although CD has no video circuitry, the first equipment used to make CD masters was video based.

For professional products, there is a need to operate at variable speed for pitch correction. When the speed of a digital recorder is reduced, the offtape sampling rate falls, and with a minimal sampling rate the first image frequency can become low enough to pass the reconstruction filter. If the sampling frequency is raised to 48 kHz without changing the response of the filters, the speed can be reduced without this problem. The currently available DVTR formats offer only 48 kHz audio sampling and so this is the only sampling rate practicable for video installations.

For landlines to FM stereo broadcast transmitters having a 15 kHz audio bandwidth, the sampling rate of 32 kHz is more than adequate, and has been in use for some time in the United Kingdom and Japan. This frequency is also in use in the NICAM 728 stereo television sound system and in DAB.

3.12 Sampling clock jitter

Figure 3.24 shows the effect of sampling clock jitter on a sloping waveform. Samples are taken at the wrong times. When these samples have passed through a system, the timebase correction stage prior to the digital-to-analog converter (DAC) will remove the jitter, and the result is shown at (b). The magnitude of the unwanted signal is proportional to the slope of the audio waveform and so the amount of jitter which can be tolerated falls at 6 dB per octave. Figure 3.25 shows the effect of differing amounts of random jitter with respect to the noise floor of various wordlengths. Note that even small amounts of jitter can degrade a 20 bit converter to the performance of a good 16 bit unit.

The allowable jitter is measured in picoseconds (ps), as shown in Figure 3.24 and clearly steps must be taken to eliminate it by design. Converter clocks must be generated from clean power supplies which are well decoupled from the power used by the logic because a converter clock must have a signal-to-noise ratio of the same order as that of the audio.

If an external clock source is used, it cannot be used directly, but must be fed through a well-designed, well-damped phase locked loop which will filter out the jitter. The phase locked loop must be built to a higher accuracy standard than in most applications.

Although it has been documented for many years, attention to control of clock jitter is not as great in actual hardware as it might be. It accounts for much of the slight audible differences between converters reproducing the same data. A well-engineered converter should substantially reject jitter on an external clock and should sound the same when reproducing the same data irrespective of the source of the data. A remote converter which sounds different after changing the type of digital cable feeding it is a dud. Unfortunately many external DACs fall into this category, as the steps outlined above have not been taken.

3.13 Aperture effect

The reconstruction process of Figure 3.22 only operates exactly as shown if the impulses are of negligible duration. In many DACs this is not the case, and many keep the analog output constant for a substantial part of the sample period or even until a different sample value is input. This produces a waveform which is more like a staircase than a pulse train. The case where the pulses have been extended in width to become equal to the sample period is known as a zero-order hold system and has a 100% aperture ratio.

Whereas pulses of negligible width have a uniform spectrum, which is flat within the audio band, pulses of 100% aperture ratio have a $\sin x/x$ spectrum which is shown in Figure 3.26. The frequency response falls to a null at the sampling rate, and as a result is about 4 dB down at the edge of the audio band. An appropriate equalization circuit can render the overall response flat once more. An alternative is to use resampling, which is shown in Figure 3.27. Resampling

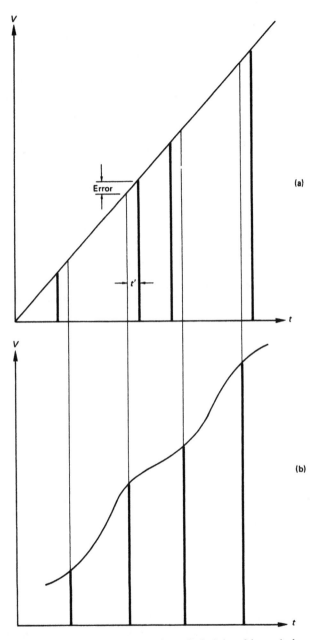

Figure 3.24 The effect of sampling timing jitter on noise, and calculation of the required accuracy for a 16-bit system. (a) Ramp sampled with jitter has error proportional to slope. (b) When jitter is removed by later circuits, error appears as noise added to samples. For a 16-bit system there are $2^{16}Q$, and the maximum slope at 20 kHz will be $20\,000\pi \times 2^{16}\ Q$ per second. If jitter is to be neglected, the noise must be less than $\tfrac{1}{2}Q$, thus timing accuracy t' multiplied by maximum slope = $\tfrac{1}{2}Q$ or $20\,000\pi \times 2^{16}Qt' = \tfrac{1}{2}Q$

$$\therefore t' = \frac{1}{2 \times 20\,000 \times \pi \times 2^{16}} = 121\ \text{ps}$$

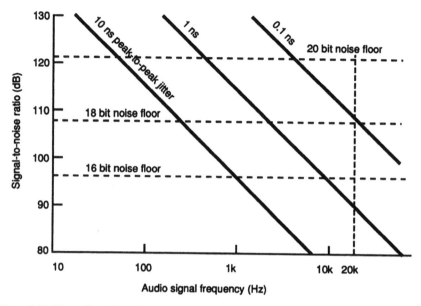

Figure 3.25 Effects of sample clock jitter on signal-to-noise ratio at different frequencies, compared with theoretical noise floors of systems with different resolutions. (After W. T. Shelton, with permission.)

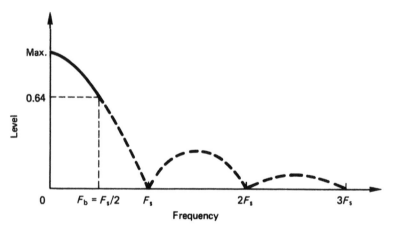

Figure 3.26 Frequency response with 100% aperture nulls at multiples of sampling rate. Area of interest is up to half sampling rate.

passes the zero-order hold waveform through a further synchronous sampling stage which consists of an analog switch which closes briefly in the centre of each sample period. The output of the switch will be pulses which are narrower than the original. If, for example, the aperture ratio is reduced to 50% of the sample period, the first frequency response null is now at twice the sampling rate, and the loss at the edge of the audio band is reduced. As the figure shows, the frequency response becomes flatter as the aperture ratio falls. The process should

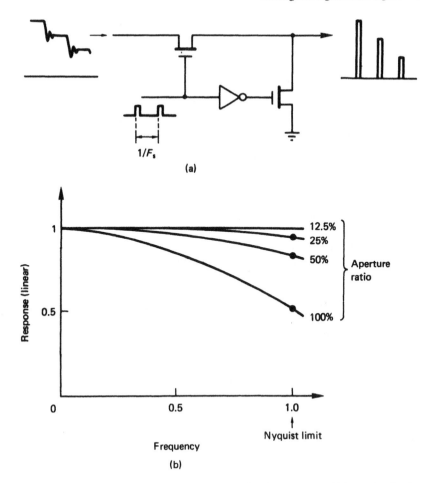

Figure 3.27 Resampling circuit eliminates transients and reduces aperture ratio. (b) Response of various aperture ratios.

not be carried too far, as with very small aperture ratios there is little energy in the pulses and noise can be a problem. A practical limit is around 12.5% where the frequency response is virtually ideal.

3.14 Quantizing

Quantizing is the process of expressing some infinitely variable quantity by discrete or stepped values (Figure 3.28). In audio the values to be quantized are infinitely variable voltages from an analog source. Strict quantizing is a process which operates in the voltage domain only.

Figure 3.29(a) shows that the process of quantizing divides the voltage range up into quantizing intervals Q, also referred to as steps S. In digital audio the quantizing intervals are made as identical as possible so that the binary numbers are truly proportional to the original analog voltage. Then the digital equivalents

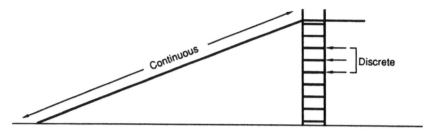

Figure 3.28 An analog parameter is continuous, whereas a quantized parameter is restricted to certain values. Here the sloping side of a ramp can be used to obtain any height, whereas a ladder only allows discrete heights.

of mixing and gain changing can be performed by adding and multiplying sample values.

Whatever the exact voltage of the input signal, the quantizer will locate the quantizing interval in which it lies. In what may be considered a separate step, the quantizing interval is then allocated a code value which is typically some form of binary number. The information sent is the number of the quantizing interval in which the input voltage lay. Whereabouts that voltage lay within the interval is not conveyed, and this mechanism puts a limit on the accuracy of the quantizer. When the number of the quantizing interval is converted back to the analog domain, it will result in a voltage at the centre of the quantizing interval as this minimizes the magnitude of the error between input and output. The number range is limited by the wordlength of the binary numbers used. In a 16-bit system there are 65 536 different quantizing intervals.

3.15 Quantizing error

It is possible to draw the transfer function for an ideal quantizer followed by an ideal DAC, and this is also shown in Figure 3.29. A transfer function is simply a graph of the output with respect to the input. In audio, when the term linearity is used, this generally means the straightness of the transfer function. Linearity is a goal in audio, yet it will be seen that an ideal quantizer is anything but linear.

Figure 3.29(b) shows the transfer function is somewhat like a staircase, and zero volts analog, corresponding to all zeros digital or muting, is halfway up a quantizing interval, or on the centre of a tread. This is the so-called mid-tread quantizer which is universally used in audio.

Quantizing causes a voltage error in the audio sample which is given by the difference between the actual staircase transfer function and the ideal straight line. This is shown in Figure 3.29(d) to be a sawtooth-like function which is periodic in Q. The amplitude cannot exceed $\pm\frac{1}{2}Q$ peak-to-peak unless the input is so large that clipping occurs.

As the transfer function is non-linear, ideal quantizing can cause distortion which produces harmonics. Unfortunately these harmonics are generated *after* the anti-aliasing filter, and so any which exceed half the sampling rate will alias. Figure 3.30 shows how this results in anharmonic distortion within the audio band. These anharmonics result in spurious tones known as birdsinging. When

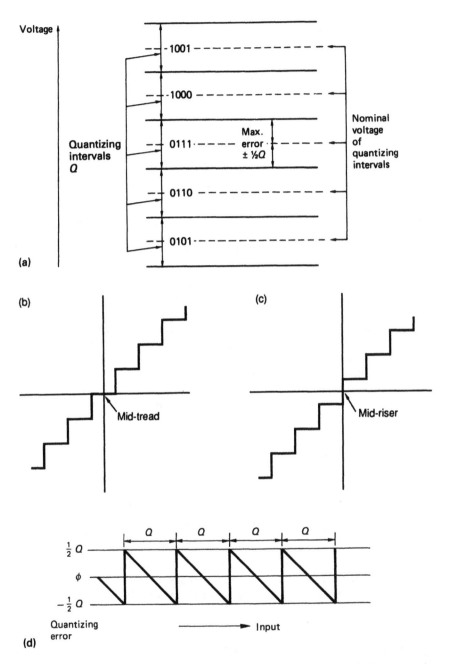

Figure 3.29 Quantizing assigns discrete numbers to variable voltages. All voltages within the same quantizing interval are assigned the same number which causes a DAC to produce the voltage at the centre of the intervals shown by the dashed lines in (a). This is the characteristic of the mid-tread quantizer shown in (b). An alternative system is the mid-riser system shown in (c). Here 0 volts analog falls between two codes and there is no code for zero. Such quantizing cannot be used prior to signal processing because the number is no longer proportional to the voltage. Quantizing error cannot exceed $\pm\frac{1}{2}Q$ as shown in (d).

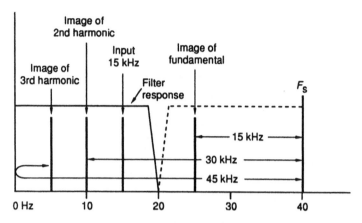

Figure 3.30 Quantizing produces distortion *after* the anti-aliasing filter; thus the distortion products will fold back to produce anharmonics in the audio band. Here the fundamental of 15 kHz produces second and third harmonic distortion at 30 and 45 kHz. This results in aliased products at $40 - 30 = (-)5$ kHz.

the sampling rate is a multiple of the input frequency the result is harmonic distortion. Where more than one frequency is present in the input, intermodulation distortion occurs, which is known as granulation.

Any one of the above effects would preclude the use of an ideal quantizer for high quality work. Fortunately they can be eliminated completely in practical equipment by the use of dither.

At high signal levels, quantizing error is effectively noise. As the audio level falls, the quantizing error of an ideal quantizer becomes more strongly correlated with the signal and the result is distortion. If the quantizing error can be decorrelated from the input in some way, the system can remain linear but noisy. Dither performs the job of decorrelation by making the action of the quantizer unpredictable and gives the system a noise floor like an analog system.

All practical digital audio systems use non-subtractive dither where the dither signal is added prior to quantization and no attempt is made to remove it at the DAC [3.5]. The introduction of dither prior to a conventional quantizer inevitably causes a slight reduction in the signal-to-noise ratio attainable, but this reduction is a small price to pay for the elimination of non-linearities.

The addition of dither means that successive samples effectively find the quantizing intervals in different places on the voltage scale. The quantizing error becomes a function of the dither, rather than a predictable function of the input signal. The quantizing error is not eliminated, but the subjectively unacceptable distortion is converted into a broadband noise which is more benign to the ear.

3.16 Requantizing and digital dither

Advanced ADC technology allows 18- and 20-bit resolution to be obtained, with perhaps more in the future. The situation then arises that an existing 16-bit device such as a digital recorder needs to be connected to the output of an ADC with greater wordlength. The words need to be shortened in some way.

Shortening the wordlength of a sample reduces the number of quantizing intervals available without changing the signal amplitude. As Figure 3.31 shows, the quantizing intervals become larger and the original signal is *requantized* with the new interval structure. This will introduce requantizing distortion having the same characteristics as quantizing distortion in an ADC. It is then obvious that when shortening, say, the wordlength of a 20-bit converter to 16 bits, the four low order bits must be removed in a way that displays the same overall quantizing structure as if the original converter had been only of 16-bit wordlength. It will be seen from Figure 3.31 that truncation cannot be used because it does not meet the above requirement but results in signal-dependent offsets because it always rounds in the same direction. Proper numerical rounding is essential in audio applications.

Requantizing by numerical rounding accurately simulates analog quantizing to the new interval size. Unfortunately the 20-bit converter will have a dither amplitude appropriate to quantizing intervals one-sixteenth the size of a 16-bit unit and the result will be highly non-linear.

In practice, the wordlength of samples must be shortened in such a way that the requantizing error is converted to noise rather than distortion. One technique which meets this requirement is to use digital dithering [3.6] prior to rounding. This is directly equivalent to the analog dithering in an ADC. Digital dither is a pseudo-random sequence of numbers. If it is required to simulate analog dither then the noise must be bipolar so that it can have an average voltage of zero. Two's complement coding must be used for the dither values as it is for the audio samples.

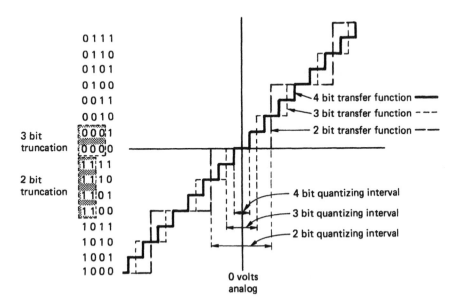

Figure 3.31 Shortening the wordlength of a sample reduces the number of codes which can describe the voltage of the waveform. This makes the quantizing steps bigger, hence the term requantizing. It can be seen that simple truncation or omission of the bits does not give analogous behaviour. Rounding is necessary to give the same result as if the larger steps had been used in the original conversion.

Figure 3.32 shows a simple digital dithering system for shortening sample wordlength. The output of a two's complement pseudo-random sequence generator of appropriate wordlength is added to input samples prior to rounding. The most significant of the bits to be discarded is examined in order to determine whether the bits to be removed sum to more or less than half a quantizing interval. The dithered sample is either rounded down, i.e. the unwanted bits are simply discarded, or rounded up, i.e. the unwanted bits are discarded but one is added to the value of the new short word. The rounding process is no longer deterministic because of the added dither which provides a linearizing random component.

3.17 Two's complement coding

In the two's complement system, the upper half of the pure binary number range has been redefined to represent negative quantities. If a pure binary counter is constantly incremented and allowed to overflow, it will produce all the numbers in the range permitted by the number of available bits, and these are shown for a 4-bit example drawn around the circle in Figure 3.33. In two's complement, the quantizing range represented by the circle of numbers does not start at zero, but starts on the diametrically opposite side of the circle. Zero is mid-range, and all numbers with the MSB (most significant bit) set are considered negative. The MSB is thus the equivalent of a sign bit where 1 = minus. Two's complement notation differs from pure binary in that the most significant bit is inverted in order to achieve the half-circle rotation.

Figure 3.34 shows how a real ADC is configured to produce two's complement output. At (a) an analog offset voltage equal to one-half the quantizing range is added to the bipolar analog signal in order to make it unipolar as at (b).

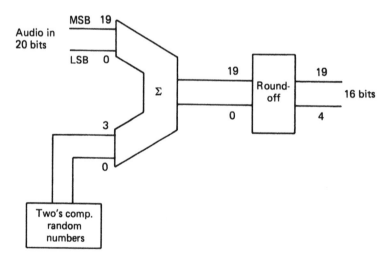

Figure 3.32 In a simple digital dithering system, two's complement values from a random number generator are added to low-order bits of the input. The dithered values are then rounded up or down according to the value of the bits to be removed. The dither linearizes the requantizing.

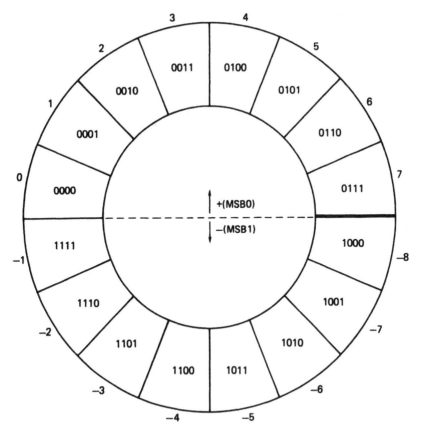

Figure 3.33 In this example of a 4-bit two's complement code, the number range is from −8 to +7. Note that the MSB determines polarity.

The ADC produces positive only numbers at (c) which are proportional to the input voltage. The MSB is then inverted at (d) so that the all-zeros code moves to the centre of the quantizing range.

Figure 3.35 shows how the two's complement system allows two sample values to be added, or mixed in audio parlance, in a manner analogous to adding analog signals in an operational amplifier. The waveform of input A is depicted by solid black samples, and that of B by samples with a solid outline. The result of mixing is the linear sum of the two waveforms obtained by adding pairs of sample values. The dashed lines depict the output values. Beneath each set of samples is the calculation which will be seen to give the correct result. Note that the calculations are pure binary. No special arithmetic is needed to handle two's complement numbers.

It is often necessary to phase reverse or invert an audio signal, for example a microphone input to a mixer. The process of inversion in two's complement is simple. All bits of the sample value are inverted to form the one's complement, and one is added. This can be checked by mentally inverting some of the values in Figure 3.33. The inversion is transparent and performing a second inversion gives the original sample values. Using inversion, signal subtraction can be

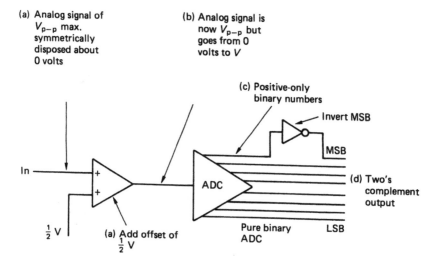

Figure 3.34 A two's complement ADC. At (a) an analog offset voltage equal to one-half the quantizing range is added to the bipolar analog signal in order to make it unipolar as at (b). The ADC produces positive-only numbers at (c), but the MSB is then inverted at (d) to give a two's complement output.

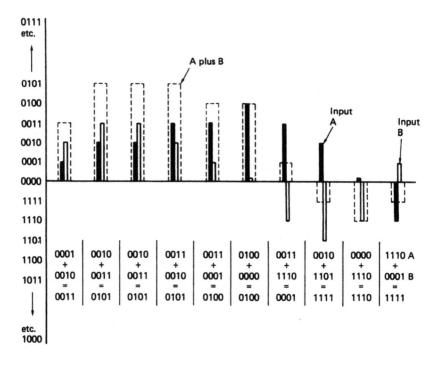

Figure 3.35 Using two's complement arithmetic, single values from two waveforms are added together with respect to midrange to give a correct mixing function.

performed using only adding logic. The inverted input is added to perform a subtraction, just as in the analog domain.

3.18 Level in digital audio

Analog tape recorders use operating levels which are some way below saturation. The range between the operating level and saturation is called the headroom. In this range, distortion becomes progressively worse and sustained recording in the headroom is avoided. However, transients may be recorded in the headroom as the ear cannot respond to distortion products unless they are sustained. The PPM level meter has an attack time constant which simulates the temporal distortion sensitivity of the ear. If a transient is too brief to deflect a PPM into the headroom, it will not be heard either.

Operating levels are used in two ways. On making a recording from a microphone, the gain is increased until distortion is just avoided, thereby obtaining a recording having the best signal-to-noise ratio (SNR). In post production the gain will be set to whatever level is required to obtain the desired subjective effect in the context of the programme material. This is particularly important to broadcasters who require the relative loudness of different material to be controlled so that the listener does not need to make continuous adjustments to the volume control.

In order to maintain level accuracy, analog recordings are traditionally preceded by line-up tones at standard operating level. These are used to adjust the gain in various stages of dubbing and transfer along land lines so that no level changes occur to the programme material.

Figure 3.36 shows some audio waveforms at various levels with respect to the coding values. Where an audio waveform just fits into the quantizing range without clipping, it has a level which is defined as 0 dBFs where Fs indicates *full scale*. Reducing the level by 6.02 dB makes the signal half as large and results in the second bit in the sample becoming the same as the sign bit. Reducing the level by a further 6.02 dB to −12 dBFs will make the second and third bits the same as the sign bit and so on.

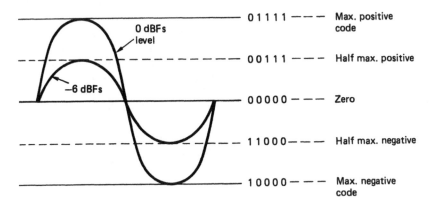

Figure 3.36 0 dBFs is defined as the level of the largest sinusoid which will fit into the quantizing range without clipping.

Unlike analog recorders, digital recorders do not have headroom, as there is no progressive onset of distortion until converter clipping, the equivalent of saturation, occurs at 0 dBFs. Accordingly many digital recorders have level meters which read in dBFs. The scales are marked with 0 at the clipping level and all operating levels are below that. This causes no difficulty provided the user is aware of the consequences.

However, in the situation where a digital copy of an analog tape is to be made, it is very easy to set the input gain of the digital recorder so that line-up tone from the analog tape reads 0 dB. This lines up digital clipping with the analog operating level. When the tape is dubbed, all signals in the headroom suffer converter clipping.

In order to prevent such problems, manufacturers and broadcasters have introduced artificial headroom on digital level meters, simply by calibrating the scale and changing the analog input sensitivity so that 0 dB analog is some way below clipping. Unfortunately there has been little agreement on how much artificial headroom should be provided, and machines which have it are seldom labelled with the amount. There is an argument which suggests that the amount of headroom should be a function of the sample wordlength, but this causes difficulties when transferring from one wordlength to another. The EBU [3.7] concluded that a single relationship between analog and digital level was desirable. In 16-bit working, 12 dB of headroom is a useful figure, but now that 18- and 20-bit converters are available, the new EBU draft recommendation specifies 18 dB.

3.19 The AES/EBU interface

The AES/EBU digital audio interface, originally published in 1985 [3.8], was proposed to embrace all the functions of existing formats in one standard. The goal was to ensure interconnection of professional digital audio equipment irrespective of origin. Alongside the professional format, Sony and Philips developed a similar format now known as SPDIF (Sony Philips Digital Interface) intended for consumer use. This offers different facilities to suit the application, yet retains sufficient compatibility with the professional interface so that, for many purposes, consumer and professional machines can be connected together [3.9, 3.10].

It was desired to use the existing analog audio cabling in such installations, which would be 600 Ω balanced line screened, with one cable per audio channel, or in some cases one twisted pair per channel with a common screen. At audio frequency the impedance of cable is high and the 600 Ω figure is that of the source and termination. If a single serial channel is to be used, the interconnect has to be self-clocking and self-synchronizing, i.e. the single signal must carry enough information to allow the boundaries between individual bits, words and blocks to be detected reliably. To fulfil these requirements, the AES/EBU and SPDIF interfaces use FM channel code which is DC-free, strongly self-clocking and capable of working with a changing sampling rate. FM code generation is described in Section 4.14. Synchronization of the deserialization process is achieved by violating the usual encoding rules.

The use of FM means that the channel frequency is the same as the bit rate when sending data ones. Tests showed that in typical analog audio-cabling installations, sufficient bandwidth was available to convey two digital audio channels in one twisted pair. The standard driver and receiver chips for RS-422A [3.11] data communication (or the equivalent CCITT-V.11) are employed for professional use, but work by the BBC [3.12] suggested that equalization and transformer coupling are desirable for longer cable runs, particularly if several twisted pairs occupy a common shield. Successful transmission up to 350 m has been achieved with these techniques [3.13]. Figure 3.37 shows the standard configuration. The output impedance of the drivers will be about 110 Ω, and the impedance of the cable used should be similar at the frequencies of interest. The driver was specified in AES-3-1985 to produce between 3 and 10 V p–p into such an impedance but this was changed to between 2 and 7 V in AES-3-1992 to reflect better the characteristics of actual RS-422 driver chips.

The original receiver impedance was set at a high 250 Ω, with the intention that up to four receivers could be driven from one source. This has been found to be inadvisable on long cables because of reflections caused by impedance mismatches and AES-3-1992 is now a point-to-point interface with source, cable and load impedance all set at 110 Ωs.

In Figure 3.38, the specification of the receiver is shown in terms of the minimum eye pattern which can be detected without error. It will be noted that the voltage of 200 mV specifies the height of the eye opening at a width of half a channel bit period. The actual signal amplitude will need to be larger than this, and even

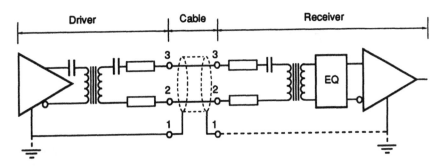

Figure 3.37 Recommended electrical circuit for use with the standard two-channel interface.

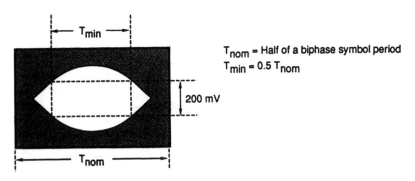

T_{nom} = Half of a biphase symbol period
T_{min} = 0.5 T_{nom}

200 mV

Figure 3.38 The minimum eye pattern acceptable for correct decoding of standard two-channel data.

larger if the signal contains noise. Figure 3.39 shows the recommended equalization characteristic which can be applied to signals received over long lines.

The purpose of the standard is to allow the use of existing analog cabling, and as an adequate connector in the shape of the XLR is already in wide service, the connector made to IEC 268 Part 12 has been adopted for digital audio use. Effectively, existing analog audio cables having XLR connectors can be used without alteration for digital connections. The AES/EBU standard does, however, require that suitable labelling should be used so that it is clear that the connections on a particular unit are digital.

The need to drive long cables does not generally arise in the domestic environment, and so a low-impedance balanced signal is not necessary. The electrical interface of the consumer format uses a 0.5 V peak single-ended signal, which can be conveyed down conventional audio-grade coaxial cable connected with RCA 'phono' plugs. Figure 3.40 shows the resulting consumer interface as specified by IEC 958.

There is a separate proposal [3.14] for a professional interface using coaxial cable for distances of around 1000 m. This is simply the AES/EBU protocol but with a 75 Ω coaxial cable carrying a 1 V signal so that it can be handled by ana-

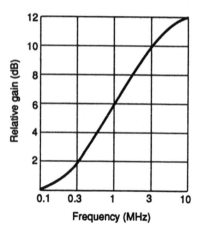

Figure 3.39 EQ characteristic recommended by the AES to improve reception in the case of long lines.

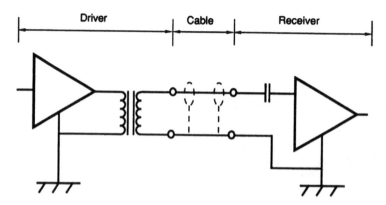

Figure 3.40 The consumer electrical interface.

log video distribution amplifiers. Impedance converting transformers allow balanced 110 Ω to unbalanced 75 Ω matching.

In Figure 3.41 the basic structure of the professional and consumer formats can be seen. One subframe consists of 32 bit-cells, of which four will be used by a synchronizing pattern. Subframes from the two audio channels, A and B, alternate on a time division basis. Up to 24-bit sample wordlength can be used, which should cater for all conceivable future developments, but normally 20-bit maximum length samples will be available with four auxiliary data bits, which can be used for a voice-grade channel in a professional application. In a consumer RDAT machine, subcode can be transmitted in bits 4–11, and the 16-bit audio in bits 12–27.

Preceding formats sent the most significant bit (MSB) first. Since this was the order in which bits were available in successive approximation converters it has become a de facto standard for inter-chip transmission inside equipment. In contrast, this format sends the least significant bit first. One advantage of this approach is that simple serial arithmetic is then possible on the samples because the carries produced by the operation on a given bit can be delayed by a one-bit period and then included in the operation on the next higher-order bit.

The format specifies that audio data must be in two's complement coding. Whilst pure binary could accept various alignments of different wordlengths with only a level change, this is not true of two's complement. If different wordlengths are used, the MSBs must always be in the same bit position otherwise the polarity will be misinterpreted. Thus the MSB has to be in bit 27 irrespective of wordlength. Shorter words are leading zero filled up to the 20-bit capacity. The channel status data included from AES-3-1992 signalling of the actual audio wordlength used so that receiving devices could adjust the digital dithering level needed to shorten a received word which is too long or pack samples onto a disk more efficiently.

Four status bits accompany each subframe. The validity flag will be reset if the associated sample is reliable. Whilst there have been many aspirations regarding what the V-bit could be used for, in practice a single bit cannot specify much, and if combined with other V-bits to make a word, the time resolution is lost. AES-3-1992 described the V-bit as indicating that the information in the associated subframe is 'suitable for conversion to an analog signal'. Thus it might be reset if the interface was being used for non-audio data as is done, for example, in CD-I players.

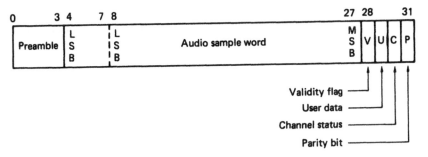

Figure 3.41 The basic subframe structure of the AES/EBU format. Sample can be 20 bits with 4 auxiliary bits, or 24 bits. LSB is transmitted first.

The parity bit produces even parity over the subframe, such that the total number of ones in the subframe is even. This allows for simple detection of an odd number of bits in error, but its main purpose is that it makes successive sync patterns having the same polarity, which can be used to improve the probability of detection of sync. The user and channel-status bits are discussed later.

Two of the subframes described above make one frame, which repeats at the sampling rate in use. The first subframe will contain the sample from channel A, or from the left channel in stereo working. The second subframe will contain the sample from channel B, or the right channel in stereo. At 48 kHz, the bit rate will be 3.072 MHz, but as the sampling rate can vary, the clock rate will vary in proportion.

In order to separate the audio channels on receipt, the synchronizing patterns for the two subframes are different as Figure 3.42 shows. These sync patterns begin with a run length of 1.5 bits which violates the FM channel coding rules and so cannot occur due to any data combination. The type of sync pattern is denoted by the position of the second transition which can be 0.5, 1.0 or 1.5 bits away from the first. The third transition is designed to make the sync patterns DC free.

The channel status and user bits in each subframe form serial data streams with 1-bit of each per audio channel per frame. The channel status bits are given a block structure and synchronized every 192 frames, which at 48 kHz gives a block rate of 250 Hz, corresponding to a period of 4 ms. In order to synchronize the channel-status blocks, the channel A sync pattern is replaced for one frame only by a third sync pattern which is also shown in Figure 3.42. The AES standard refers to these as X,Y and Z, whereas IEC 958 calls them M,W and B. As stated, there is a parity bit in each subframe, which means that the binary level at the end of a subframe will always be the same as at the beginning. Since the sync patterns have the same characteristic, the effect is that sync patterns always have

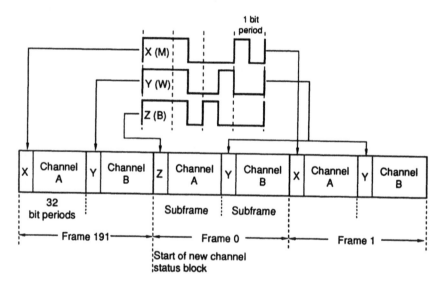

Figure 3.42 Three different preambles (X, Y and Z) are used to synchronize a receiver at the starts of subframes.

the same polarity and the receiver can use that information to reject noise. The polarity of transmission is not specified, and indeed an accidental inversion in a twisted pair is of no consequence, since it is only the transition that is of importance, not the direction.

When 24-bit resolution is not required, which is most of the time, the four auxiliary bits can be used to provide voice co-ordination between studios as well as programme exchange on the same cables. Twelve-bit samples of the talkback signal are taken at one-third the main sampling rate. Each 12-bit sample is then split into three nibbles (half a byte, for gastronomers) which can be sent in the auxiliary data slot of three successive samples in the same audio channel. As there are 192 nibbles per channel status block period, there will be exactly 64 talkback samples in that period. The reassembly of the nibbles can be synchronized by the channel status sync pattern as shown in Figure 3.43. Channel status byte 2 reflects the use of auxiliary data in this way.

In both the professional and consumer formats, the sequence of channel status bits over 192 subframes builds up a 24-byte channel status block. However, the contents of the channel status data are completely different between the two applications. The professional channel status structure is shown in Figure 3.44. Byte 0 determines the use of emphasis and the sampling rate, with details in Figure 3.45. Byte 1 determines the channel usage mode, i.e. whether the data transmitted are a stereo pair, two unrelated mono signals or a single mono signal, and details the user bit handling. Figure 3.46 gives details. Byte 2 determines wordlength as in Figure 3.47. This was made more comprehensive in AES-3-1992. Byte 3 is applicable only to multi-channel applications. Byte 4 indicates the suitability of the signal as a sampling rate reference and will be discussed in more detail later in this chapter.

There are two slots of four bytes each which are used for alphanumeric source and destination codes. These can be used for routing. The bytes contain 7-bit ASCII characters (printable characters only) sent least significant bit (LSB) first, with the eighth bit set to zero according to AES-3-1992. The destination code can be used to operate an automatic router, and the source code will allow the origin of the audio and other remarks to be displayed at the destination.

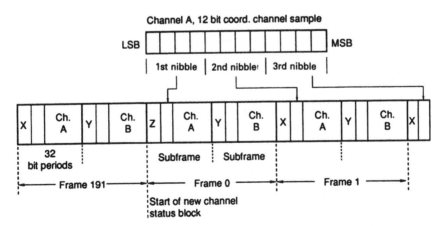

Figure 3.43 The coordination signal is of a lower bit rate to the main audio and thus may be inserted in the auxiliary nibble of the interface subframe, taking three subframes per coordination sample.

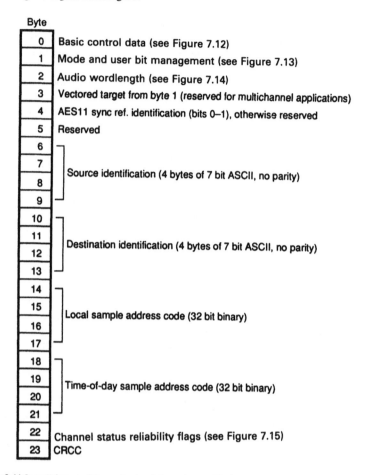

Byte

0	Basic control data (see Figure 7.12)
1	Mode and user bit management (see Figure 7.13)
2	Audio wordlength (see Figure 7.14)
3	Vectored target from byte 1 (reserved for multichannel applications)
4	AES11 sync ref. identification (bits 0–1), otherwise reserved
5	Reserved
6	
7	Source identification (4 bytes of 7 bit ASCII, no parity)
8	
9	
10	
11	Destination identification (4 bytes of 7 bit ASCII, no parity)
12	
13	
14	
15	Local sample address code (32 bit binary)
16	
17	
18	
19	Time-of-day sample address code (32 bit binary)
20	
21	
22	Channel status reliability flags (see Figure 7.15)
23	CRCC

Figure 3.44 Overall format of the professional channel status block.

Bytes 14–17 convey a 32-bit sample address which increments every channel status frame. It effectively numbers the samples in a relative manner from an arbitrary starting point. Bytes 18–21 convey a similar number, but this is a time-of-day count, which starts from zero at midnight. As many digital audio devices do not have real time clocks built in, this cannot be relied upon. AES-3-92 specified that the time-of-day bytes should convey the real time at which a recording was made, making it rather like timecode. There are enough combinations in 32 bits to allow a sample count over 24 hours at 48 kHz. The sample count has the advantage that it is universal and independent of local supply frequency. In theory if the sampling rate is known, conventional hours, minutes, seconds, frames timecode can be calculated from the sample count, but in practice it is a lengthy computation and users have proposed alternative formats in which the data from EBU or SMPTE timecode is transmitted directly in these bytes. Some of these proposals are in service as de facto standards.

The penultimate byte contains four flags which indicate that certain sections of the channel status information are unreliable (Figure 3.48). This allows the

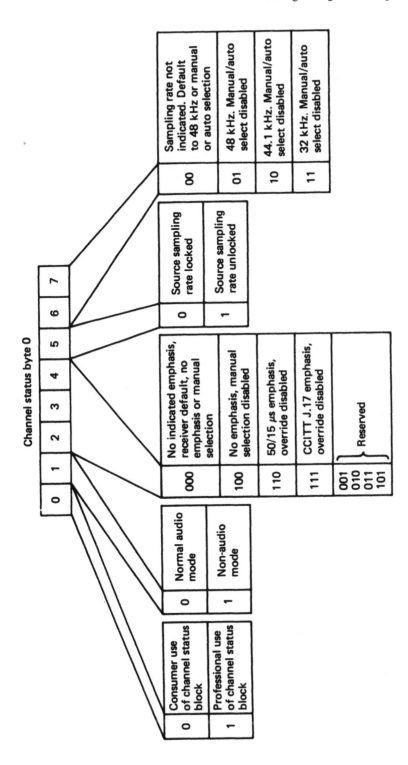

Figure 3.45 The first byte of the channel status information in the AES/EBU standard deals primarily with emphasis and sampling-rate control.

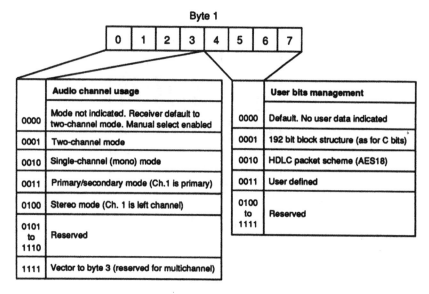

Figure 3.46 Format of byte 1 of professional channel status.

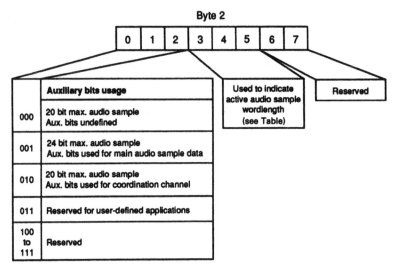

Bits states 3 4 5	Audio wordlength (24 bit mode)	Audio wordlength (20 bit mode)
0 0 0	Not indicated	Not indicated
0 0 1	23 bits	19 bits
0 1 0	22 bits	18 bits
0 1 1	21 bits	17 bits
1 0 0	20 bits	16 bits
1 0 1	24 bits	20 bits

Figure 3.47 Format of byte 2 of professional channel status.

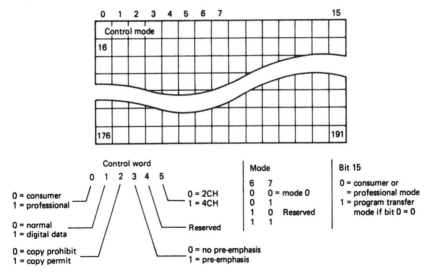

Figure 3.48 Byte 22 of channel status indicates if some of the information in the block is unreliable.

transmission of an incomplete channel status block where the entire structure is not needed or where the information is not available. For example, setting bit 5 to a logical one would mean that no origin or destination data would be interpreted by the receiver, and so it need not be sent.

The final byte in the message is a CRCC which converts the entire channel-status block into a codeword. The channel status message takes 4 ms at 48 kHz and in this time a router could have switched to another signal source. This would damage the transmission, but will also result in a CRCC failure so the corrupt block is not used. Error correction is not necessary, as the channel status data are either stationary, i.e. they stay the same, or change at a predictable rate, e.g. timecode. Stationary data will only change at the receiver if a good CRCC is obtained.

For consumer use, a different version of the channel status specification is used which is incompatible with the professional structure. The first bit of the consumer channel status message is set to zero so that professional machines do not attempt to interpret it erroneously.

References

3.1 Martin, W.H. Decibel – the name for the transmission unit. *Bell Syst. Tech. J.* (Jan. 1929)

3.2 Meyer, J. Time correction of anti-aliasing filters used in digital audio systems. *J. Audio Eng. Soc.*, **32**, 132–137 (1984)

3.3 Lipshitz, S.P. Pockock, M. and Vanderkooy, J., On the audibility of midrange phase distortion in audio systems. *J. Audio Eng. Soc.*, **30**, 580–595 (1982)

3.4 Preis, D. and Bloom, P.J. Perception of phase distortion in anti-alias filters. *J. Audio Eng. Soc.*, **32**, 842–848 (1984)

3.5 Vanderkooy, J. and Lipshitz, S.P. Resolution below the least significant bit in digital systems with dither. *J. Audio Eng. Soc.*, **32**, 106–113 (1984)

3.6   Vanderkooy, J. and Lipshitz, S.P. Digital dither. *81st Audio Eng. Soc.* Conv. preprint 2412 (C-8). Los Angeles: Audio Eng. Soc (1986)

3.7   Moller, L. Signal levels across the EBU/AES digital audio interface. *Proc. 1st NAB Radio Montreux Symp.*, 16–28. Montreux

3.8   Audio Engineering Society AES recommended practice for digital audio engineering – serial transmission format for linearly represented digital audio data. *J. Audio Eng. Soc.* **33**, 975–984 (1985)

3.9   EIAJ CP-340. *A Digital Audio Interface.* Tokyo: EIAJ (1987)

3.10 EIAJ CP-1201. *Digital Audio Interface (revised). Tokyo: EIAJ (1992)*

3.11 EIA RS-422A. Electronic Industries Association, 2001 Eye St N.W., Washington, DC 20006, USA

3.12 Smart, D.L., Transmission performance of digital audio serial interface on audio tie lines. *BBC Designs Dept Techn. Memo.,* 3.296/84

3.13 European Broadcasting Union. Specification of the digital audio interface. *EBU Doc. Tech.,* 3250

3.14 Rorden, B. and Graham, M. A proposal for integrating digital audio distribution into TV production. *J. SMPTE,* 606–608 (Sept.1992)

Chapter 4

Audio recording

4.1 Introduction

There are now more ways than ever to record audio for television production. VTRs, both analog and digital, have audio tracks, but with the use of timecode, audio only recorders can be synchronized so that they operate at the same speed as a video recorder. Synchronized audio recording can be carried out on tape, or using hard disks; a technology which has become increasingly popular because of the speed of operation it allows. As all of these techniques rely on magnetism, this topic must be considered the starting point of any discussion of recording [4.1].

4.2 Magnetism

A magnetic field can be created by passing a current through a solenoid, which is no more than a coil of wire. When the current ceases, the magnetism disappears. However, many materials, some quite common, display a permanent magnetic field with no apparent power source. Magnetism of this kind results from the spin of electrons within atoms. Atomic theory describes atoms as having nuclei around which electrons orbit, spinning as they go. Different orbits can hold a different number of electrons. The distribution of electrons determines whether the element is diamagnetic (non-magnetic) or paramagnetic (magnetic characteristics are possible). Diamagnetic materials have an even number of electrons in each orbit but half of them spin in each direction. The opposed spins cancel any resultant magnetic moment. Fortunately there are certain elements, the transition elements, which have an odd number of electrons in certain orbits. The magnetic moment due to electronic spin is not cancelled out in these paramagnetic materials.

Figure 4.1 shows that paramagnetism materials can be classified as antiferromagnetic, ferrimagnetic and ferromagnetic. In some materials alternate atoms are antiparallel and so the magnetic moments are cancelled. In ferrimagnetic materials there is a certain amount of antiparallel cancellation, but a net magnetic moment remains. In ferromagnetic materials such as iron, cobalt or nickel, all of the electron spins can be aligned and as a result the most powerful magnetic behaviour is obtained.

It is not immediately clear how a material in which electron spins are parallel could ever exist in an unmagnetized state or how it could be partially magnetized by a relatively small external field. The theory of magnetic domains has been

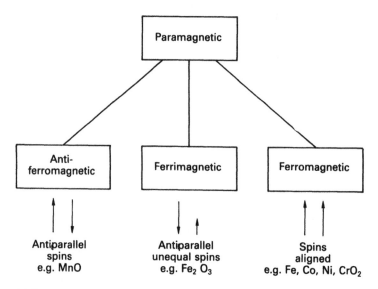

Figure 4.1 The classification of paramagnetic materials. The ferromagnetic materials exhibit the strongest magnetic behaviour.

developed to explain what is observed in practice. Figure 4.2(a) shows a ferromagnetic bar which is demagnetized. It has no net magnetic moment because it is divided into domains or volumes which have equal and opposite moments. Ferromagnetic material divides into domains in order to reduce its magnetostatic energy. Figure 4.2(b) shows a domain wall which is around 0.1 μm thick. Within the wall the axis of spin gradually rotates from one state to another. An external field of quite small value is capable of disturbing the equilibrium of the domain wall by favouring one axis of spin over the other. The result is that the domain wall moves and one domain becomes larger at the expense of another. In this way the net magnetic moment of the bar is no longer zero as shown in (c).

For small distances, the domain wall motion is linear and reversible if the change in the applied field is reversed. However, larger movements are irreversible because heat is dissipated as the wall jumps to reduce its energy. Following such a domain wall jump, the material remains magnetized after the external field is removed and an opposing external field must be applied which must do further work to bring the domain wall back again. This is a process of hysteresis where work must be done to move each way. Were it not for this non-linear mechanism, magnetic recording would be impossible. If magnetic materials were linear, tapes would return to the demagnetized state immediately after leaving the field of the head and this chapter would not have much to say.

Figure 4.3 shows a hysteresis loop which is obtained by plotting the magnetization M when the external field H is swept to and fro. On the macroscopic scale, the loop appears to be a smooth curve, whereas on a small scale it is in fact composed of a large number of small jumps. These were first discovered by Barkhausen. Starting from the unmagnetized state at the origin, as an external field is applied, the response is initially linear and the slope is given by the susceptibility. As the applied field is increased a point is reached where the

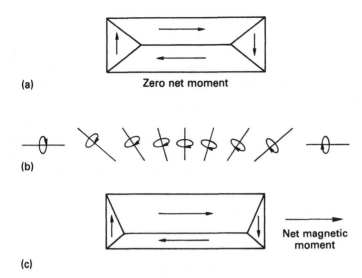

Figure 4.2 (a) A magnetic material can have a zero net moment if it is divided into domains as shown here. Domain walls (b) are areas in which the magnetic spin gradually changes from one domain to another. The stresses which result store energy. When some domains dominate, a net magnetic moment can exist as in (c).

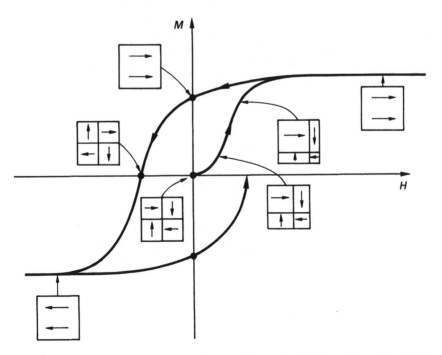

Figure 4.3 A hysteresis loop which comes about because of the non-linear behaviour of magnetic materials. If this characteristic were absent, magnetic recording would not exist.

magnetization ceases to increase. This is the saturation magnetization M_s. If the applied field is removed, the magnetization falls, not to zero, but the remanent magnetization M_r. This remanence is the magnetic memory mechanism which makes recording possible. The ratio of M_r to M_s is called the squareness ratio. In recording media, squareness is beneficial as it increases the remanent magnetization.

If an increasing external field is applied in the opposite direction, the curve continues to the point where the magnetization is zero. The field required to achieve this is called the intrinsic coercive force $_mH_c$. A small increase in the reverse field reaches the point where, if the field where to be removed, the remanent magnetization would become zero. The field required to do this is the remanent coercive force, $_rH_c$.

As the external field H is swept to and fro, the magnetization describes a major hysteresis loop. Domain wall transit causes heat to be dissipated on every cycle around the loop and the dissipation is proportional to the loop area. For a recording medium, a large loop is beneficial because the replay signal is a function of the remanence and high coercivity resists erasure. Heating is not an issue. For a device such as a recording head, a small loop is beneficial. Figure 4.4(a) shows the large loop of a hard magnetic material used for recording media and for permanent magnets. Figure 4.4(b) shows the small loop of a soft magnetic material which is used for recording heads and transformers.

According to the Nyquist noise theorem, anything which dissipates energy when electrical power is supplied must generate a noise voltage when in thermal equilibrium. Thus magnetic recording heads have a noise mechanism which is due to their hysteretic behaviour. The smaller the loop, the less the hysteretic noise. In conventional heads, there are a large number of domains and many small domain wall jumps. In thin film heads there are fewer domains and the jumps must be larger. The noise this causes is known as Barkhausen noise, but as the same mechanism is responsible it is not possible to say at what point hysteresis noise should be called Barkhausen noise.

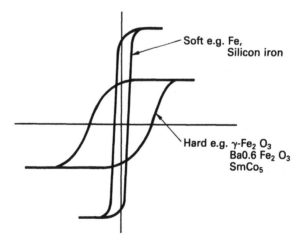

Figure 4.4 The recording medium requires a large loop area (a), whereas the head requires a small loop area (b) to cut losses.

4.3 Magnetic recording

Magnetic recording relies on the hysteresis of magnetically hard media as described above. By definition the transfer function is non-linear, and so the recorded signal will be a distorted version of the input. Analog magnetic recorders have to use bias to linearize the process, whereas digital recorders are not concerned with the non-linearity, and bias is unnecessary.

As heads and media cannot know the meaning of the waveforms which they handle, they cannot be analog or digital; instead the term describes the way in which the replayed signals are *interpreted*. When the replay process simply outputs the varying waveform from the tape, the machine is analog. When discrete decisions are made on the replay signal, the player is digital.

The signal-to-noise ratio (SNR) of an analog tape track is effectively the SNR of the recording, whereas in a digital recorder it is only necessary to distinguish between a few levels and a much poorer track SNR is allowable. The SNR of a tape track depends on the width. Doubling the width doubles the amplitude of the replay signal, but in the case of the noise it is the power which is doubled because it does not add coherently. Consequently there is a 6 dB increase in signal level, but only a 3 dB increase in noise level. Thus analog recorders have wide tracks, whereas digital recorders have narrow tracks.

Figure 4.5 shows the construction of a magnetic tape record head. Heads designed for use with tape work in actual contact with the magnetic coating. The tape is tensioned to pull it against the head. There will be a wear mechanism and need for periodic cleaning. A magnetic circuit carries a coil through which the record current passes and generates flux. A non-magnetic gap forces the flux to leave the magnetic circuit of the head and penetrate the medium. The most efficient recording will be obtained when the reluctance of the magnetic circuit is dominated by that of the gap. This means making the ring structure only just large enough to fit the coil to shorten the magnetic circuit as much as possible.

Using thin-film heads, the magnetic circuits and windings are produced by deposition on a substrate at right angles to the tape plane, and as seen in Figure 4.6 they can be made very accurately at small track spacings. Perhaps more importantly, because the magnetic circuits do not have such large parallel areas,

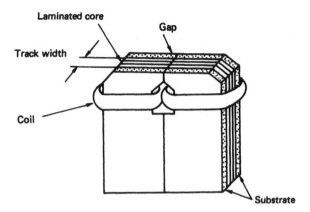

Figure 4.5 A digital record head is similar in principle to an analog head but uses much narrower tracks.

mutual inductance and crosstalk are smaller, allowing a higher practical track density.

Figure 4.7 shows that the recording is actually made just after the trailing pole of the record head where the flux strength from the gap is falling. The width of the gap is generally made quite large to ensure that the full thickness of the magnetic coating is recorded.

A conventional inductive head has a frequency response, shown in Figure 4.8. At 0 Hz there is no change of flux and no output. As a result inductive heads are at a disadvantage at very low speeds. The output rises with frequency until the rise is halted by the onset of thickness loss. As the frequency rises, the recorded wavelength falls and flux from the shorter magnetic patterns cannot be picked up so far away. At some point, the wavelength becomes so short that flux from the back of the tape coating cannot reach the head and a decreasing thickness of tape contributes to the replay signal [4.2]. In digital recorders using short wavelengths to obtain high density, there is no point in using thick coatings.

As wavelength further reduces, gap loss occurs, where the head gap is too big to resolve detail on the track. The result is an aperture effect (see Chapter 3) where the response has nulls as flux cancellation takes place across the gap. Clearly the smaller the gap the shorter the wavelength of the first null. This

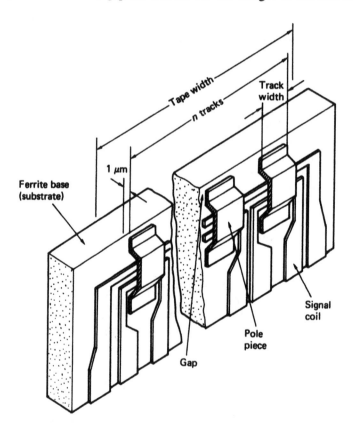

Figure 4.6 The thin-film head shown here can be produced photographically with very small dimensions. Flat structure reduces crosstalk and allows a finer track pitch to be used.

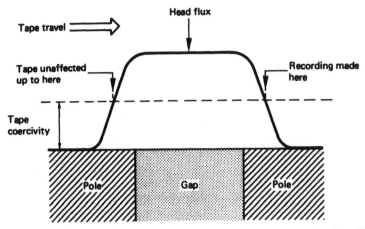

Figure 4.7 The recording is actually made near the trailing pole of the head where the head flux falls below the coercivity of the tape.

contradicts the requirement of the record head to have a large gap. In quality analog recorders, it is the norm to have different record and replay heads for this reason, and the same will be true in digital machines which have separate record and playback heads. Clearly, where the same pair of heads are used for record and play, the head gap size will be determined by the playback requirement. It will also be seen from Figure 4.8 that there are some irregularities in the low frequency response region which are known as head bumps. These irregularities come about because of interactions between longer wavelengths on the medium and the width of the head poles. Figure 4.8 shows the response of a digital head in which bumps are less of a problem.

Analog audio recorders must use physically wide heads to drive the head bumps down to low frequencies, although at high tape speeds, such as 30 inches per second (i.p.s.), this is impossible and the head bumps become audible. Large head poles result in low contact pressure. In digital recorders, channel coding can be used to narrow the recorded spectrum, and the heads can be made much smaller with a corresponding increase in contact pressure over the gap. Thin film heads necessarily have very small poles and this results in the head bumps moving up the band.

As can be seen, the frequency response of a magnetic recorder is far from ideal, and both analog and digital recorders must use some form of equalization.

4.4 Biased analog recording

Figure 4.9 shows a typical analog tape deck [4.3]. There are three heads – erase, record and replay – and means to drive the tape at constant speed and to wind or rewind it so that appropriate points in the recording can be located. The tape may be kept on open reels or in a cassette. The erase head erases a wider path on the tape than the track width to ensure that new recordings are always made in fully erased areas even if there is slight head misalignment. Similarly the replay head is slightly smaller than the track width to prevent lateral alignment variation altering the amplitude of the replay signal.

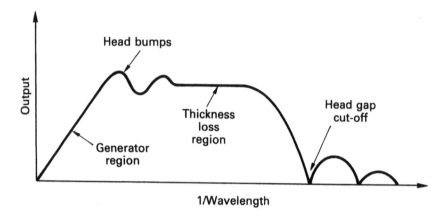

Figure 4.8 The frequency response of a conventional inductive head. See text for details.

The hysteresis which makes magnetic recording possible unfortunately results in distortion if some steps are not taken to compensate [4.4]. In analog recording a high frequency sine wave signal known as bias is used. Figure 4.10 shows that such a signal produces a magnetic field strength which diminishes with distance away from the head gap. A particle of tape coating experiences a waveform decaying with respect to time as it recedes from the trailing pole of the head. In the case of an erase head, only the high frequency oscillation is supplied. As the tape recedes from the head it will be swept around a set of diminishing hysteresis loops until it is demagnetized. The same technique is used when demagnetizing heads. A degausser is a solenoid connected to the AC mains supply via a switch. An alternating field from a degausser is applied, and the degausser is physically removed to a distance before being switched off, resulting in exactly the same decaying waveform.

In the case of the record head, the bias is linearly added to the audio waveform to be recorded. Now as the tape recedes from the trailing pole it will experience a decaying oscillation superimposed upon the audio signal. Figure 4.11 shows that the result is that the tape is swept through a series of minor loops which converge upon the magnetization determined by the audio waveform, linearizing the transfer function.

There is no optimum amplitude of the bias signal and a practical compromise must be made between distortion, replay level and frequency response which will need to be individually assessed for each type of tape used. Often the bias level is increased until the replay output has reduced by 2–3 dB. In very short wavelength recording, such as in compact cassettes, high audio frequencies can add to the bias and reduce the signal level. This is known as HF squashing. In the Dolby HX (headroom extension) system, the HF content of the input audio is monitored and used to modulate the bias level.

The presence of a significant bias level also affects the maximum level which can be recorded. Distortion begins when the level of signal plus bias reaches tape saturation. Consequently analog tape has a maximum operating level (MOL) often defined as the level above which distortion exceeds 0.1%. Gross distortion occurs at saturation and between MOL and saturation is a region known as

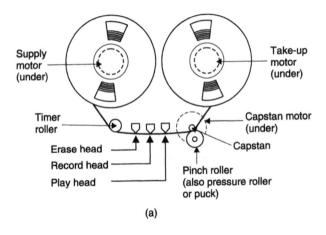

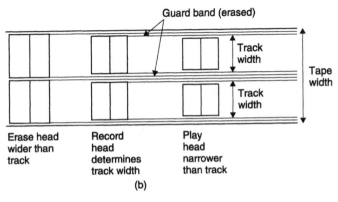

Figure 4.9 (a) An analog tape deck. The capstan and pressure roller provides constant tape speed and the supply and take-up reels apply tape tension. The supply reel back tension is obtained by driving the motor lightly against its rotation causes the tape to be held against the heads. (b) The heads have different track widths to ensure that slight vertical misalignment does not affect recording or replay.

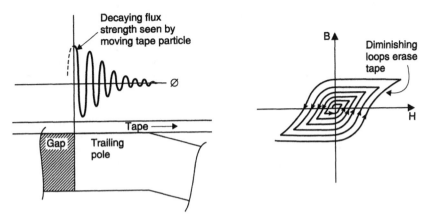

Figure 4.10 A high frequency signal produces a field which appears to decay with time to a tape receding from the gap. This sweeps the tape around a set of diminishing loops until it is erased.

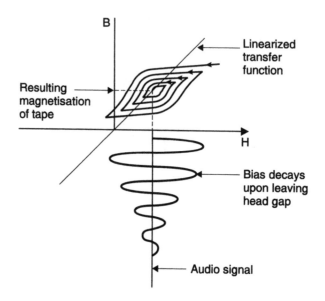

Figure 4.11 When bias is added to the audio signal the tape is swept through a series of loops which converge upon a straight transfer function.

headroom in which transients can be recorded because the hearing mechanism cannot register the distortion unless it is sustained.

4.5 Head alignment

Figure 4.12 shows the three axes in which a head can be aligned. If the *wrap* adjustment is incorrect, the maximum pressure due to the tape tension will not act at the gap and high frequency response will be impaired. If the *zenith* angle is incorrect, tape tension will be greater at one edge than the other, leading to wear problems and a differing frequency responses in analog multi-track machines.

The azimuth angle is critical because if it is incorrect it makes the effective head gap larger, increasing the aperture effect of the head and impairing HF response. It will be clear from Figure 4.13 that azimuth error is more significant in the wide tracks of professional analog recorders than it is with the narrow tracks of a digital recorder.

Azimuth is adjusted using a reference tape containing transitions recorded at 90 degrees to the tape edge. The replay head is adjusted for maximum output on a high frequency, or if the head has more than one track it may be adjusted to align the phases of two replay signals on a dual trace oscilloscope. Caution is needed to ensure that alignment is not achieved with exactly one cycle of phase error.

The record head azimuth is adjusted using a blank tape which is played back at the now adjusted replay head.

In stereo analog machines, it is important that the record azimuth adjustment is not made when the bias levels in the two channels are unknown. As recording actually takes place at some point after the trailing pole where the bias field

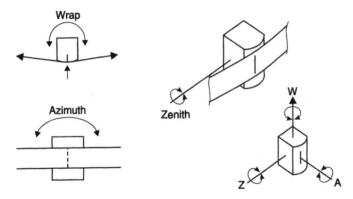

Figure 4.12 The three axes in which heads can be aligned. Wrap centres the tape tension pressure on the gap. Zenith equalizes the pressure across the tape and azimuth sets the gap at 90° to the tape path.

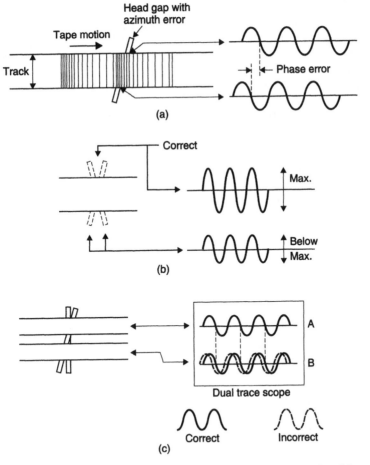

Figure 4.13 (a) Azimuth error has a greater effect on wide tracks. (b) Azimuth can be adjusted for maximum output with a high frequency test tape or (c) on a stereo machine for minimum phase error between the tracks.

decays, a different bias level in the two channels has the same effect as an azimuth error. Where a machine is being extensively overhauled, the azimuth may need to be adjusted before a bias adjustment can be performed. In this case it will be necessary to repeat the record azimuth adjustment.

4.6 Pre-emphasis and equalization

Both of these are processes in which frequency dependent filtering is applied. They differ in principle and achieve different purposes. Pre-emphasis is part of the interchange standard for the tape format and is designed to reduce high frequency noise by boosting HF in the record process and applying an equal and opposite response on playback. Equalization (eq) is necessary to compensate for losses and deficiencies in the record and playback processes. In the absence of pre-emphasis the actual magnetization of the tape should be constant with respect to frequency. Consequently any departure from a flat frequency response in the record process is compensated by record eq, and any departure in the replay process is similarly corrected by replay eq. These compensations are highly machine dependent and vary with head wear. Thus, eq is adjustable, whereas pre-emphasis is fixed. Different tape formats use different pre-emphasis, and a switch may be provided to select the correct pre-emphasis so that one machine may be used with several formats.

Figure 4.14 shows the pre-emphasis and the adjustable record eq which is used to compensate for HF losses in the record head. The replay circuits contain adjustable eq to counteract the effects of Figure 4.8 as well as fixed de-emphasis.

As the record and replay equalizations are in series, then in theory it would be possible to make all of the corrections on, say, replay. However, although the machine would be able to play its own tapes, these tapes would display response irregularities when played elsewhere. Similarly tapes from elsewhere would have an irregular response when played on such a machine. Thus it is the requirements of interchange which demand separate record and replay eq.

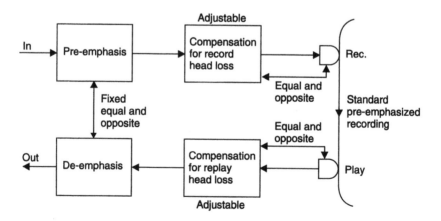

Figure 4.14 The record and replay equalization is needed to compensate for head losses to give a signal on tape which deviates from flat only by the standard pre-emphasis curve.

The replay eq is adjusted using a test tape with the correct pre-emphasis containing tones of equal level at a range of frequencies. The eq is adjusted so that all of the tones play back at the same level. The record eq is then adjusted by recording onto a blank tape which uses the replay system as a reference.

Care should be taken at LF that one of the test tones does not coincide with a head bump as this could result in a false adjustment. Where head bumps are suspected, LF eq should be adjusted to give a good average response over a range of low frequencies.

The main disadvantage of high quality analog recorders is that they have to be cleaned and adjusted frequently in order to maintain performance. As there are so many adjustments this is quite a labour-intensive process and in a commercial enterprise can be costly.

4.7 Analog noise reduction

The random noise caused by the finite number of particles in tape limits the performance of analog recorders. The only way in which the noise can actually be reduced is to make the tracks wider, but only 3 dB improvement is obtained for every doubling of track width, so this is not a practical solution. As the noise of tape recording is significantly higher in level than noise from any other component in an analog production system, efforts have been made to reduce its effect. Of these the best known and most successful have been the various systems produced by Dolby laboratories [4.5].

Noise reducers (NR) are essentially reversible analog compressors and function by compressing the dynamic range of the signal at the encoder prior to recording in a way which can be reversed at the decoder on replay. The expansion process reduces the effect of tape noise.

As noise reduction is designed to work with unmodified analog recorders the decoder must work on the audio signal only. Figure 4.15 shows that this is overcome by using a recursive type compressor which operates based on its own output. Clearly when this is done, the decoder and the encoder can be made complementary because both work from the same signal, provided the recorder has no gain error and a reasonably flat frequency response. Gain error is overcome by generating a NR line-up tone in the encoder which is put on tape before the programme. On replay the input level of the decoder is adjusted during the line-up tone to give a particular level indication.

In compression system the gain variation could result in the noise floor rising and falling. If audible this will result in annoying breathing or pumping effects. The noise floor will be audible whenever it is not masked. Consequently practical NR systems must divide the audio spectrum into sub-bands in which the gain is individually controlled. The performance of this approach increases with the number of bands. In the Dolby A system the audio input is divided into four bands at 80 Hz, 3 kHz and 9 kHz.

The Dolby SR system has a novel approach in which the sub-band boundaries can move. The spectrum of the signal is analysed and a sub-band boundary is positioned just above the significant energy in the signal. This allows the maximum masking effect.

Correct replay of an NR encoded tape can only be achieved with the correct decoder. Consequently it is important that tapes are marked with the type of NR

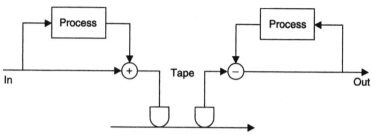

Tape = In + Process × In = In (1 + Process)
Out = Tape − Process × Out
∴ Tape = Out + Process × Out = (1 + Process)
∴ In (1 + Process) = Out (1 + Process)
∴ In = Out

Figure 4.15 In order to obtain complementary expansion of the compressed signal, noise reducers work by analysing their own output. The decoder then has access to the same signal as the encoder (assuming the tape deck has unity gain).

used. A drawback of analog NR is that errors in replay level caused by dropouts are expanded. High quality professional analog tapes are specifically designed to minimize dropout frequency.

4.8 Digital magnetic recording

In digital recording, the amplitude of the record current is constant, and recording is performed by reversing the direction of the current with respect to time. As the track passes the head, this is converted to the reversal of the magnetic field left on the tape with respect to distance. The magnetic recording is therefore bipolar [4.6].

The record current is selected to be as large as possible to obtain the best SNR without resulting in transition spreading which occurs as saturation is approached. In practice, the best value for the record current may be that which minimizes the error rate.

Figure 4.16 shows what happens when a conventional inductive head, i.e. one having a normal winding, is used to replay the bipolar track made by reversing the record current. The head output is proportional to the rate of change of flux and so only occurs at flux reversals. In other words, the replay head differentiates the flux on the track.

The amplitude of the replay signal is of no consequence and often an automatic gain control (AGC) system is used. What matters is the time at which the write current, and hence the flux stored on the medium, reverses. This can be determined by locating the peaks of the replay impulses, which can conveniently be done by differentiating the signal and looking for zero crossings. Figure 4.17 shows that this results in noise between the peaks. This problem is overcome by the gated peak detector, where only zero crossings from a pulse which exceeds the threshold will be counted. The AGC system allows the thresholds to be fixed. As an alternative, the record waveform can also be restored by integration, which opposes the differentiation of the head as in Figure 4.18 [4.7].

A more recent development is the magneto-resistive (M-R) head. This is a head

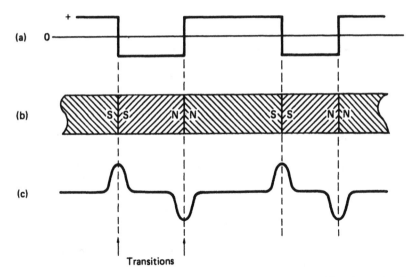

Figure 4.16 Basic digital recording. In (a) the write current in the head is reversed from time to time, leaving a binary magnetization pattern shown in (b). When replayed, the waveform in (c) results because an output is only produced when flux in the head changes. Changes are referred to as transitions.

which measures the flux on the tape rather than using it to generate a signal directly. Flux measurement works down to DC and so offers advantages at low tape speeds. Recorders which have low head-to-medium speed, such as DCC (digital compact cassette), tend use M-R heads, whereas recorders with high speeds, such as digital VTRs, DAT and magnetic disk drives tend to use inductive heads.

Digital recorders are sold into a highly competitive market and must operate at high density in order to be commercially viable. As a result the shortest possible wavelengths will be used. Figure 4.19 shows that when two flux changes, or transitions, are recorded close together, they affect each other on replay. The amplitude of the composite signal is reduced, and the position of the peaks is pushed outwards. This is known as inter-symbol interference, or peak-shift distortion, and it occurs in all magnetic media. The effect is primarily due to high frequency loss and it can be reduced by eq on replay.

In digital recording the goal is not to reproduce the recorded waveform, but to reproduce the data it represents. Small irregularities are of no consequence, especially as an error correction system will be fitted. Consequently digital recorders have fewer adjustments. Azimuth adjustment is generally only necessary if a head is replaced, sometimes not even then. There is no bias to adjust and some machines automatically adjust the replay eq to optimize the error rate. Digital machines require much less maintenance and, if the tape is in a cassette, much less operator skill.

4.9 Time compression

When samples are converted, the ADC must run at a constant clock rate and it outputs an unbroken stream of samples. Time compression allows the sample stream to be broken into blocks for convenient handling.

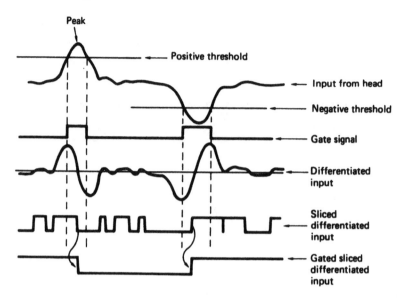

Figure 4.17 Gated peak detection rejects noise by disabling the differentiated output between transitions.

Figure 4.20 shows an ADC feeding a pair of RAMs. When one is being writ-ten by the ADC, the other can be read, and vice-versa. As soon as the first RAM is full, the ADC output switches to the input of the other RAM so that there is no loss of samples. The first RAM can then be read at a higher clock rate than the sampling rate. As a result the RAM is read in less time than it took to write it, and the output from the system then pauses until the second RAM is full. The samples are now time compressed. Instead of being an unbroken stream which is difficult to handle, the samples are now arranged in blocks with convenient

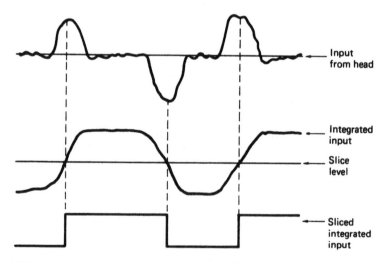

Figure 4.18 Integration method for re-creating write-current waveform.

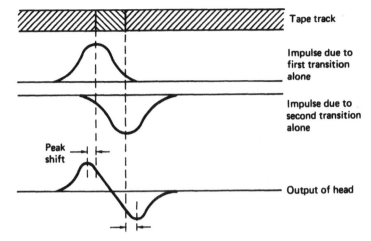

Figure 4.19 Readout pulses from two closely recorded transitions are summed in the head and the effect is that the peaks of the waveform are moved outwards. This is known as peak-shift distortion and equalization is necessary to reduce the effect.

pauses in between them. In these pauses numerous processes can take place. A rotary head recorder might switch heads; a hard disk might move to another track. On a tape recording, the time compression of the audio samples allows time for synchronizing patterns, subcode and error correction words to be recorded. In digital video recorders time compression allows the continuous audio samples to be placed in blocks recorded at the same bit rate as the video data.

Subsequently, any time compression can be reversed by time expansion. Samples are written into a RAM at the incoming clock rate, but read out at the standard sampling rate. Unless there is a design fault, time compression is totally inaudible. In a recorder, the time expansion stage can be combined with the timebase correction stage so that speed variations in the medium can be eliminated at the same time. The use of time compression is universal in digital audio recording. In general the *instantaneous* data rate at the medium is not the same as the rate at the converters, although clearly the *average* rate must be the same.

Another application of time compression is to allow more than one channel of audio to be carried on a single cable. In the AES/EBU interface, for example, audio samples are time compressed by a factor of two, so it is possible to carry samples from a stereo source in one cable.

4.10 Practical digital recorders

As the process of timebase correction can be used to eliminate irregularities in data flow, digital recorders do not need to record continuously as analog recorders do and so there is much more design freedom. Recorders can be made which have discontinuous tracks or tracks which are subdivided into blocks. Rotary head recorders and disk drives have many advantages over the stationary head approach of the traditional analog tape recorder.

In a rotary head recorder, the heads are mounted in a revolving drum and the tape is wrapped around the surface of the drum in a helix as can be seen in

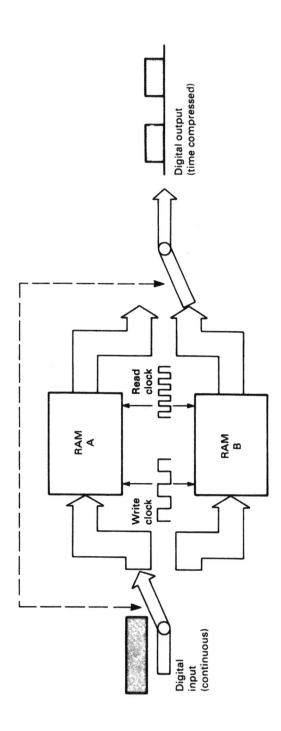

Figure 4.20 In time compression, the unbroken real-time stream of samples from an ADC is broken up into discrete blocks. This is accomplished by the configuration shown here. Samples are written into one RAM at the sampling rate by the write clock. When the first RAM is full, the switches change over, and writing continues into the second RAM whilst the first is read using a higher-frequency clock. The RAM is read faster than it was written and so all the data will be output before the other RAM is full. This opens spaces in the data flow which are used as described in the text.

Figure 4.21. The helical tape path results in the heads traversing the tape in a series of diagonal or slanting tracks. The space between the tracks is controlled not by head design but by the speed of the tape and in modern recorders this space is reduced to zero with corresponding improvement in packing density.

The added complexity of the rotating heads and the circuitry necessary to control them is offset by the improvement in density. The discontinuous tracks of the rotary head recorder are naturally compatible with time compressed data. As Figure 4.21 illustrates, the audio samples are time compressed into blocks each of which can be contained in one slant track.

Rotary head recording is naturally complemented by azimuth recording. Figure 4.22(a) shows that in azimuth recording, the transitions are laid down at an angle to the track by using a head which is tilted. Machines using azimuth recording must always have an even number of heads, so that adjacent tracks can be recorded with opposite azimuth angle. The two track types are usually referred to as A and B. Figure 4.22(b) shows the effect of playing a track with the wrong type of head. The playback process suffers from an enormous azimuth error. The effect of azimuth error can be understood by imagining the tape track to be made from many identical parallel strips. In the presence of azimuth error, the strips at one edge of the track are played back with a phase shift relative to strips at the other side. At some wavelengths, the phase shift will be 180 degrees, and there will be no output; at other wavelengths, especially long wavelengths, some output will reappear. The effect is rather like that of a comb filter, and serves to attenuate crosstalk due to adjacent tracks so that no guard bands are required. Since no tape is wasted between the tracks, more efficient use is made of the tape. The term guard-band-less recording is often used instead of, or in addition to, the term azimuth recording. The failure of the azimuth effect at long wavelengths is a characteristic of azimuth recording, and it is necessary to ensure

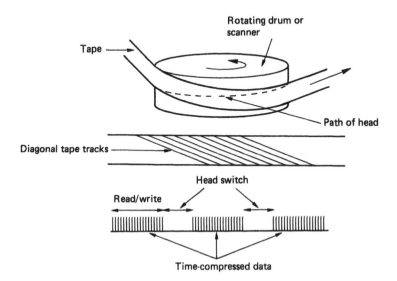

Figure 4.21 In a rotary-head recorder, the helical tape path around a rotating head results in a series of diagonal or slanting tracks across the tape. Time compression is used to create gaps in the recorded data which coincide with the switching between tracks.

that the spectrum of the signal to be recorded has a small low-frequency content. The signal will need to pass through a rotary transformer to reach the heads, and cannot therefore contain a DC component.

In recorders such as DAT [4.8] there is no separate erase process, and erasure is achieved by overwriting with a new waveform. Overwriting is only successful when there are no long wavelengths in the earlier recording, since these penetrate deeper into the tape, and the short wavelengths in a new recording will not be able to erase them. In this case the ratio between the shortest and longest wavelengths recorded on tape should be limited. Restricting the spectrum of the signal to allow erasure by overwrite also eases the design of the rotary transformer.

There are two different approaches to azimuth recording as shown in Figure 4.23. In the first method, there is no separate erase head, and in order to guarantee that no previous recording can survive a new recording, the recorded tracks can be made rather narrower than the head pole simply by reducing the linear speed of the tape so that it does not advance so far between sweeps of the rotary heads. This is shown in Figure 4.24. In DAT, for example, the head pole is 20.4 mm wide, but the tracks it records are only 13.59 mm wide. Alternatively, the record head can be the same width as the track pitch and cannot guarantee complete overwrite. A separate erase head will be necessary. The advantage of this approach is that insert editing does not leave a seriously narrowed track.

As azimuth recording rejects crosstalk, it is advantageous if the replay head is some 50% wider than the tracks. It can be seen from Figure 4.25 that there will be crosstalk from tracks at both sides of the home track, but this crosstalk is attenuated by azimuth effect. The amount by which the head overlaps the adjacent track determines the spectrum of the crosstalk, since it changes the delay in the azimuth comb-filtering effect.

More importantly, the signal-to-crosstalk ratio becomes independent of tracking error over a small range, because as the head moves to one side, the loss of

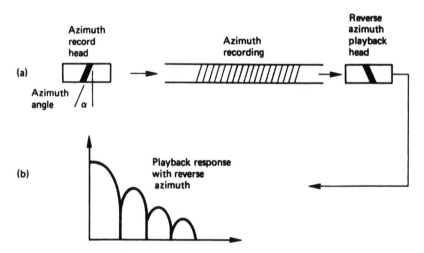

Figure 4.22 In azimuth recording (a), the head gap is tilted. If the track is played with the same head, playback is normal, but the response of the reverse azimuth head is attenuated (b).

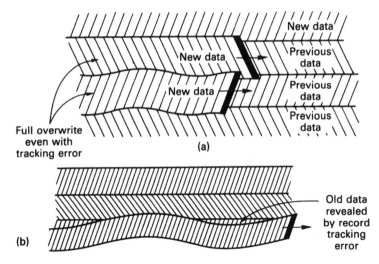

Figure 4.23 In (a) if the azimuth record head is wider than the track, full overwrite is obtained even with misalignment. In (b) if the azimuth record head is the same width as the track, misalignment results in failure to overwrite and an erase head becomes necessary.

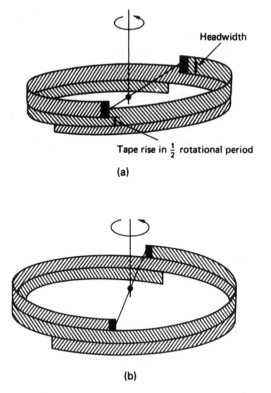

Figure 4.24 With azimuth recording, the record head may be wider than (a) or of the same width as (b) the track.

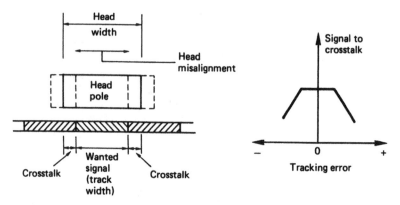

Figure 4.25 When the head pole is wider than the track, the wanted signal is picked up along with crosstalk from the adjacent tracks. If the head is misaligned, the signal-to-crosstalk ratio remains the same until the head fails to register with the whole of the wanted track.

crosstalk from one adjacent track is balanced by the increase of crosstalk from the track on the opposite side. This phenomenon allows for some loss of track straightness and for the residual error which is present in all track-following servo systems.

The hard disk recorder stores data on concentric tracks which it accesses by moving the head radially. Clearly while the head is moving it cannot transfer data. Using time compression, a hard disk drive can be made into an audio recorder with the addition of a certain amount of memory.

Figure 4.26 shows the principle. The instantaneous data rate of the disk drive is far in excess of the sampling rate at the converter, and so a large time compression factor can be used. The disk drive can read a block of data from disk, and place it in the timebase corrector in a fraction of the real time it represents in the audio waveform. As the timebase corrector steadily advances through the memory, the disk drive has time to move the heads to another track before the memory runs out of data. When there is sufficient space in the memory for another block, the drive is commanded to read, and fills up the space. Although the data transfer at the medium is highly discontinuous, the buffer memory provides an unbroken stream of samples to the DAC and so continuous audio is obtained.

Recording is performed by using the memory to assemble samples until the contents of one disk block is available. This is then transferred to disk at high data rate. The drive can then reposition the head before the next block is available in memory. An advantage of hard disks is that access to the audio is much quicker than with tape, as all of the data are available within the time taken to move the head.

Disk drives permanently sacrifice storage density in order to offer rapid access. The use of a flying head with a deliberate air gap between it and the medium is necessary because of the high medium speed, but this causes a severe separation loss which restricts the linear density available. The air gap must be accurately maintained, and consequently the head is of low mass and is mounted flexibly.

Figure 4.27 shows that the aerohydrodynamic part of the head is known as the slipper; it is designed to provide lift from the boundary layer which changes

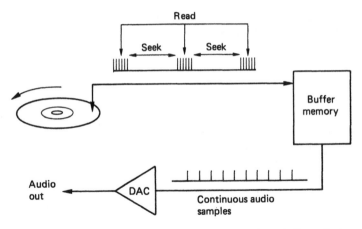

Figure 4.26 In a hard disk recorder, a large-capacity memory is used as a buffer or timebase corrector between the converters and the disk. The memory allows the converters to run constantly despite the interruptions in disk transfer caused by the head moving between tracks.

rapidly with changes in flying height. It is not initially obvious that the difficulty with disk heads is not making them fly, but making them fly close enough to the disk surface. The boundary layer travelling at the disk surface has the same speed as the disk, but as height increases, it slows down due to drag from the surrounding air. As the lift is a function of relative air speed, the closer the slipper comes to the disk, the greater the lift will be. The slipper is therefore mounted at the end of a rigid cantilever sprung towards the medium. The force with which the head is pressed towards the disk by the spring is equal to the lift at the designed flying height. Because of the spring, the head may rise and fall over small warps in the disk. It would be virtually impossible to manufacture disks flat enough to dispense with this feature. As the slipper negotiates a warp it will pitch and roll in addition to rising and falling, but it must be prevented from yawing, as this would cause an azimuth error. Downthrust is applied to the aerodynamic centre by a spherical thrust button, and the required degrees of freedom are supplied by a thin flexible gimbal.

In a moving-head device it is not practicable to position separate erase, record and playback heads accurately. Erase is by overwriting, and reading and writing

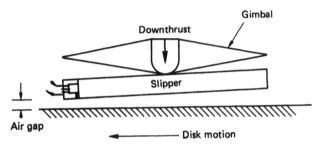

Figure 4.27 Disk head slipper develops lift from boundary layer moving with disk. This reaches equilibrium with the downthrust at the designed flying height. Resultant air gap prevents head wear but restricts storage density.

are often done by the same head. An exception is where MR read heads are used. As these cannot write, two separate heads must be closely mounted in the same slipper.

4.11 Replay synchronization

Transfer of samples between digital audio devices in real time is only possible if both use a common sampling rate and they are synchronized. A digital audio recorder must be able to synchronize to the sampling rate of a digital input in order to record the samples. It is frequently necessary for such a recorder to be able to play back locked to an external sampling rate reference so that it can be connected to, for example, a digital mixer. The process is already common in video systems but now extends to digital audio. Chapter 5 describes a digital audio reference signal (DARS).

Figure 4.28 shows how the external reference locking process works. The timebase expansion is controlled by the external reference which becomes the read clock for the RAM and so determines the rate at which the RAM address changes. In the case of a digital tape deck, the write clock for the RAM would be proportional to the tape speed. If the tape is going too fast, the write address will catch up with the read address in the memory, whereas if the tape is going too slow the read address will catch up with the write address. The tape speed is controlled by subtracting the read address from the write address. The address difference is used to control the tape speed. Thus if the tape speed is too high, the memory will fill faster than it is being emptied, and the address difference will grow larger than normal. This slows down the tape.

Thus in a digital recorder the speed of the medium is constantly changing to keep the data rate correct. Clearly this is inaudible as properly engineered timebase correction totally isolates any instabilities on the medium from the data fed to the converter.

In multi-track recorders, the various tracks can be synchronized to sample accuracy so that no timing errors can exist between the tracks. Extra transports can be slaved to the first to the same degree of accuracy if more tracks are required. In stereo recorders image shift due to phase errors is eliminated.

In order to replay without a reference, perhaps to provide an analog output, a digital recorder generates a sampling clock locally by means of a crystal oscillator. Provision will be made on professional machines to switch between internal and external references.

4.12 Error correction and concealment

In a recording of binary data, a bit is either correct or wrong, with no intermediate stage. Small amounts of noise are rejected, but inevitably, infrequent noise impulses cause some individual bits to be in error. Dropouts cause a larger number of bits in one place to be in error. An error of this kind is called a burst error. Whatever the medium and whatever the nature of the mechanism responsible, data are either recovered correctly, or suffer some combination of bit errors and burst errors. In DAT, random errors can be caused by noise, whereas burst errors are due to contamination or scratching of the tape surface.

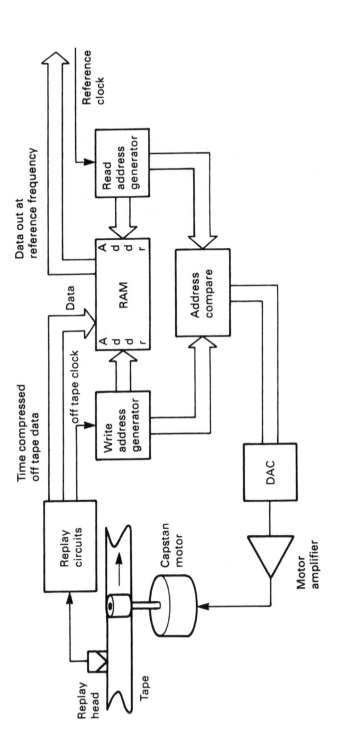

Figure 4.28 In a recorder using time compression, the samples can be returned to a continuous stream using RAM as a timebase corrector (TBC). The long-term data rate has to be the same on the input and output of the TBC or it will lose data. This is accomplished by comparing the read and write addresses and using the difference to control the tape speed. In this way the tape speed will automatically adjust to provide data as fast as the reference clock takes it from the TBC.

The audibility of a bit error depends upon which bit of the sample is involved. If the LSB of one sample was in error in a loud passage of music, the effect would be totally masked and no one could detect it. Conversely, if the MSB of one sample was in error in a quiet passage, no one could fail to notice the resulting loud transient. Clearly a means is needed to render errors from the medium inaudible. This is the purpose of error correction.

In binary, a bit has only two states. If it is wrong, it is only necessary to reverse the state and it must be right. Thus the correction process is trivial and perfect. The main difficulty is in identifying the bits which are in error. This is done by coding the data by adding redundant bits. Adding redundancy is not confined to digital technology: airliners have several engines and cars have twin braking systems. Clearly the more failures which have to be handled, the more redundancy is needed. If a four-engined airliner is designed to fly normally with one engine failed, three of the engines have enough power to reach cruise speed, and the fourth one is redundant. The amount of redundancy is equal to the amount of failure which can be handled. In the case of the failure of two engines, the plane can still fly, but it must slow down; this is graceful degradation. Clearly the chances of a two-engine failure on the same flight are remote.

In digital audio, the amount of error which can be corrected is proportional to the amount of redundancy, and within this limit, the samples are returned to exactly their original value. Consequently *corrected* samples are inaudible. If the amount of error exceeds the amount of redundancy, correction is not possible, and, in order to allow graceful degradation, concealment will be used. Concealment is a process where the value of a missing sample is estimated from those nearby. The estimated sample value is not necessarily exactly the same as the original, and so under some circumstances concealment can be audible, especially if it is frequent. However, in a well-designed system, concealments occur with negligible frequency unless there is an actual fault or problem.

Concealment is made possible by rearranging or shuffling the sample sequence prior to recording. This is shown in Figure 4.29 where odd-numbered samples are separated from even-numbered samples prior to recording. The odd and even sets of samples may be recorded in different places, so that an uncorrectable burst error only affects one set. On replay, the samples are recombined into their natural sequence, and the error is now split up so that it results in every other sample being lost. The waveform is now described half as often, but can still be reproduced with some loss of accuracy. This is better than not being reproduced at all, even if it is not perfect. Almost all digital recorders use such an odd/even shuffle for concealment. Clearly if any errors are fully correctable, the shuffle is a waste of time; it is only needed if correction is not possible.

In high density recorders, more data are lost in a given sized dropout. Adding redundancy equal to the size of a dropout to every code is inefficient. Figure 4.30(a) shows that the efficiency of the system can be raised using interleaving. Sequential samples from the ADC are assembled into codes, but these are not recorded in their natural sequence. A number of sequential codes are assembled along rows in a memory. When the memory is full, it is copied to the medium by reading down columns. On replay, the samples need to be de-interleaved to return them to their natural sequence. This is done by writing samples from tape into a memory in columns, and when it is full the memory is read in rows. Samples read from the memory are now in their original sequence so there is no

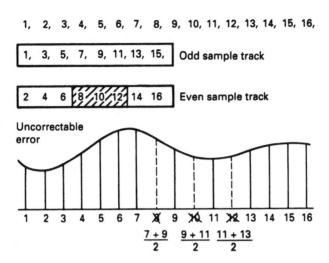

1, 2, 3, 4, 5, 6, 7, 8, 9, 10, 11, 12, 13, 14, 15, 16,

| 1, 3, 5, 7, 9, 11, 13, 15, | Odd sample track |

| 2 4 6 8 10 12 14 16 | Even sample track |

Uncorrectable error

1 2 3 4 5 6 7 8 9 10 11 12 13 14 15 16

$$\frac{7+9}{2} \quad \frac{9+11}{2} \quad \frac{11+13}{2}$$

Figure 4.29 In cases where the error correction is inadequate, concealment can be used provided that the samples have been ordered appropriately in the recording. Odd and even samples are recorded in different places as shown here. As a result an uncorrectable error causes incorrect samples to occur singly, between correct samples. In the example shown, sample 8 is incorrect, but samples 7 and 9 are unaffected and an approximation to the value of sample 8 can be had by taking the average value of the two. This interpolated value is substituted for the incorrect value.

effect on the recording. However, if a burst error occurs on the medium, it will damage sequential samples in a vertical direction in the de-interleave memory. When the memory is read, a single large error is broken down into a number of small errors whose size is exactly equal to the correcting power of the codes and the correction is performed with maximum efficiency.

An extension of the process of interleave is where the memory array has not only rows made into codewords, but also columns made into codewords by the addition of vertical redundancy. This is known as a product code [4.9]. Figure

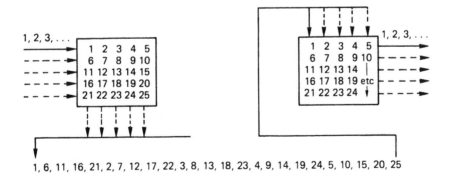

Figure 4.30(a) In interleaving, samples are recorded out of their normal sequence by taking columns from a memory which was filled in rows. On replay the process must be reversed. This puts the samples back in their regular sequence, but breaks up burst errors into many smaller errors which are more efficiently corrected. Interleaving and de-interleaving cause delay.

4.30(b) shows that in a product code the redundancy calculated first and checked last is called the outer code, and the redundancy calculated second and checked first is called the inner code. The inner code is formed along tracks on the medium. Random errors due to noise are corrected by the inner code and do not impair the burst correcting power of the outer code. Burst errors are declared uncorrectable by the inner code which flags the bad samples on the way into the de-interleave memory. The outer code reads the error flags in order to locate the erroneous data. As it does not have to compute the error locations, the outer code can correct more errors.

The interleave, de-interleave, time compression and timebase correction processes cause delay and this is evident in the time taken before audio emerges after starting a digital machine. Confidence replay takes place later than the distance between record and replay heads would indicate. In DASH format recorders, confidence replay is about one-tenth of a second behind the input. Synchronous recording requires new techniques to overcome the effect of the delays.

The presence of an error correction system means that the audio quality is independent of the tape/head quality within limits. There is no point in trying to assess the health of a machine by listening to it, as this will not reveal whether the error rate is normal or within a whisker of failure. The only useful procedure is to monitor the frequency with which errors are being corrected, and to compare it with normal figures. Professional digital audio equipment should have an error rate display.

4.13 Channel coding

In most recorders used for storing digital information, the medium carries a track which reproduces a single waveform. Clearly data words representing audio samples contain many bits and so they have to be recorded serially, a bit at a time. Recording data serially is not as simple as connecting the serial output of a shift register to the head. In digital audio, a common sample value is all zeros, as this corresponds to silence. If a shift register is loaded with all zeros and shifted out serially, the output stays at a constant low level, and nothing is recorded on the track. On replay there is nothing to indicate how many zeros were present, or even how fast to move the medium. Clearly serialized raw data cannot be recorded directly; they have to be modulated into a waveform which contains an embedded clock irrespective of the values of the bits in the samples. On replay a circuit called a data separator can lock to the embedded clock and use it to separate strings of identical bits.

The process of modulating serial data to make them self-clocking is called channel coding [4.10]. Channel coding also shapes the spectrum of the serialized waveform to make it more efficient. With a good channel code, more data can be stored on a given medium. Spectrum shaping is used in DVTRs to produce DC-free signals which will pass through rotary transformers to reach the revolving heads. It is also used in DAT to allow re-recording without erase heads. Channel coding is also needed in digital broadcasting where shaping of the spectrum is an obvious requirement to avoid interference with other services.

The FM code, also known as Manchester code or biphase mark code, shown in Figure 4.31 was the first practical self-clocking binary code and it is suitable for both transmission and recording. It is DC free and very easy to encode and

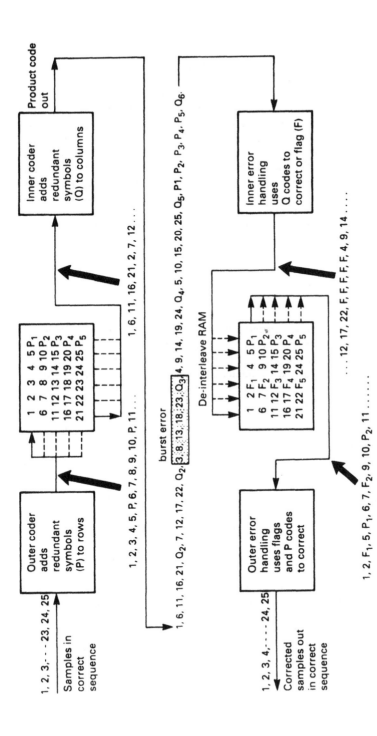

Figure 4.30(b) In addition to the redundancy P on rows, inner redundancy Q is also generated on columns. On replay, the Q code checker will pass on flags F if it finds an error too large to handle itself. The flags pass through the de-interleave process and are used by the outer error correction to identify which symbol in the row needs correcting with P redundancy. The concept of crossing two codes in this way is called a product code.

decode. It is the code specified for the AES/EBU digital audio interconnect standard described in Chapter 3. In the field of recording it remains in use today only where density is not of prime importance, for example in SMPTE/EBU timecode for professional audio and video recorders and in floppy disks.

In FM there is always a transition at the bit–cell boundary which acts as a clock. For a data one, there is an additional transition at the bit-cell centre. Figure 4.31 shows that each data bit can be represented by two channel bits. For a data zero, they will be 10, and for a data one they will be 11. Since the first bit is always one, it conveys no information, and is responsible for the density ratio of only one-half. Since there can be two transitions for each data bit, the jitter margin can only be half a bit, and the bandwidth required is high. The high clock content of FM does, however, mean that data recovery is possible over a wide range of speeds; hence the use for timecode. The lowest frequency in FM is due to a stream of zeros and is equal to half the bit rate. The highest frequency is due to a stream of ones, and is equal to the bit rate. Thus the fundamentals of FM are within a band of one octave. Effective eq is generally possible over such a band. FM is not polarity conscious and can be inverted without changing the data.

4.14 Group codes

Further improvements in coding rely on converting patterns of real data to patterns of channel bits with more desirable characteristics, using a conversion table known as a codebook. If a data symbol of m bits is considered, it can have 2^{mR} different combinations. As it is intended to discard undesirable patterns to improve the code, it follows that the number of channel bits n must be greater than m. The number of patterns which can be discarded is:

$$2^n - 2^m$$

One name for the principle is group code recording (GCR), and an important parameter is the code rate, defined as:

$$R = \frac{m}{n}$$

The amount of jitter which can be withstood is proportional to the code rate, and so a figure near to unity is desirable. The choice of patterns which are used in the codebook will be those which give the desired balance between clock content, bandwidth and DC content.

Figure 4.32 shows that the upper spectral limit can be made to be some fraction of the channel bit rate according to the minimum distance between ones in the channel bits. This is known as the minimum transition parameter. It can be obtained by multiplying the number of channel detent periods between transitions by the code rate.

The channel code of DAT will be used here as a good example of a group code. There are many channel codes available, but few of them are suitable for azimuth recording because of the large amount of crosstalk. The crosstalk cancellation of azimuth recording fails at low frequencies, so a suitable channel code must not only be free of DC, but it must suppress low frequencies as well. A further issue is that erasure is by overwriting, and as the heads are optimized for short-wavelength working, best erasure will be when the ratio between the longest and shortest wavelengths in the recording is small.

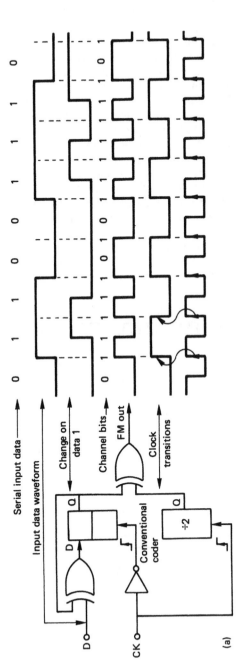

Figure 4.31 The FM waveform and the channel bits which may be used to describe transitions in it.

In Figure 4.33, some examples from the 8/10 group code of DAT are shown [4.11]. Clearly a channel waveform which spends as much time high as low has no net DC content, and so all 10-bit patterns which meet this criterion of zero disparity can be found. For every bit the channel spends high, the DC content will increase; for every bit the channel spends low, the DC content will decrease. As adjacent channel ones are permitted, the jitter margin will be 0.8 bits. Unfortunately there are not enough DC-free combinations in 10 channel bits to provide the 256 patterns necessary to record 8 data bits. A further constraint is that it is desirable to restrict the maximum run length to improve overwrite capability and reduce peak shift. In the 8/10 code of DAT, no more than three channel zeros are permitted between channel ones, which makes the longest wavelength only four times the shortest. There are only 153 10-bit patterns which are within this maximum run length and which have a digital sum value (DSV) of zero.

The remaining 103 data combinations are recorded using channel patterns that have non-zero DC content. Two channel patterns are allocated to each of the 103 data patterns. One of these has a DC value of +2, the other has a value of −2. For simplicity, the only difference between them is that the first channel bit is inverted. The choice of which channel-bit pattern to use is based on the DC due to the previous code.

For example, if several bytes have been recorded with some of the 153 DC-free patterns, the DC content of the code will be zero. The first data byte is then found which has no zero disparity pattern. If the +2 pattern is used, the code at

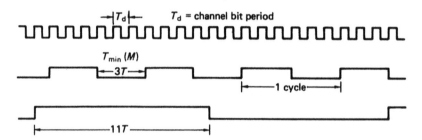

Figure 4.32 A channel code can control its spectrum by placing limits on T_{min} (M) and T_{max} which define upper and lower frequencies. The ratio of T_{max}/T_{min} determines the asymmetry of waveform and predicts DC content and peak shift. Example shown is EFM.

Eight-bit dataword	Ten-bit codeword	DSV	Alternative codeword	DSV
00010000	1101010010	0		
00010001	0100010010	2	1100010010	−2
00010010	0101010010	0		
00010011	0101110010	0		
00010100	1101110001	2	0101110001	−2
00010101	1101110011	2	0101110011	−2
00010110	1101110110	2	0101110110	−2
00010111	1101110010	0		

Figure 4.33 Some of the 8/10 codebook for non-zero DSV symbols (two entries) and zero DSV symbols (one entry).

the end of the pattern will also become +2. When the next pattern of this kind is found, the code having the value of –2 will automatically be selected to return the channel DC content to zero. In this way the code is kept DC free, but the maximum distance between transitions can be shortened. A code of this kind is known as a low disparity code. Decoding is simpler, because there is a direct relationship between 10-bit codes and 8-bit data.

4.15 Timecode synchronizing

In film and television it is frequently necessary to synchronize together two or more signal sources. An audio tape recorder must be locked to a video recorder in order to obtain lip-sync. Several audio recorders may be locked together to provide an increased number of channels.

The process requires a way of recording time alongside the audio and video signals on the media, and equipment which can read the time recordings and control tape motion. Figure 4.34 shows how linear timecode (LTC) is recorded [4.12]. The example given is for SMPTE timecode used in 60 Hz video formats, but the EBU equivalent is almost the same except that the range of the frame count is only 25. The basic data recorded are hours, minutes, seconds and video frames, expressed in a BCD (binary coded decimal) format. BCD is like binary except that a four-digit BCD number only goes from 0 to 9. Each timecode requires eight symbols, denoted HH, MM, SS, FF but some of these have a limited range, e.g. the tens of hours symbol, so not all symbols need 4 bits. Figure 4.34 shows that the data are serially recorded using FM coding as described in Section 4.13. The sync word, used for deserializing the bit stream, consists of 16 bits, of which 12 are successive ones. As the data use BCD coding, this pattern cannot occur anywhere else in the bit stream. The sync pattern is asymmetrical so that a timecode reader can establish the direction of tape motion.

In EBU timecode, the count is contiguous, but in SMPTE colour systems the frame rate is actually 29.97 Hz and so dividing by 30 would give an incorrect seconds count. The discrepancy amounts to 108 frames per hour. The solution is to have 108 virtual frames per hour which exist only to keep the seconds division process correct, but which are not actually recorded. Consequently these virtual frames cannot be found on the tape and the frame count jumps over them. In the sequence of frames from the tape, certain codes appear to have been dropped, hence the term drop frame. Provided the location of these dropped frames is standardized, no equipment will be foolish enough to search for them.

As NTSC has a two-frame colour subcarrier sequence, then in practice it is necessary drop pairs of frames to keep the same relationship between frame count and colour frame status. In order to maintain the correct seconds count very nearly correct, a pair of frames is dropped at the beginning of every minute with the exception of minutes 0, 10, 20, 30, 40 and 50. This results in an error of just over two frames per day.

In equipment which runs at exactly 60 Hz, such as certain CD mastering equipment, drop frames are unnecessary. It will be seen in Figure 4.34 that there is a flag which indicates its use. The timecode recording also has space for user data which is quite independent of the timecode function. In practice, scene and take numbers might be recorded there.

The timecode waveform to be recorded is DC free and requires only modest

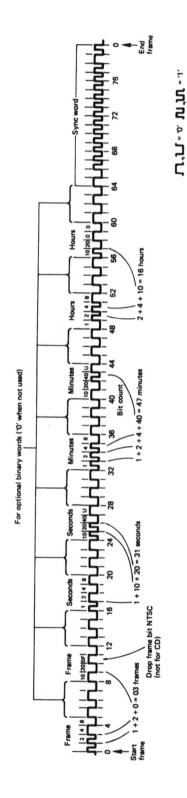

Figure 4.34 In SMPTE standard timecode, the frame number and time are stored as eight BCD symbols. There is also space for 32 user-defined bits. The code repeats every frame. Note the asymmetrical sync word which allows the direction of tape movement to be determined.

bandwidth by analog audio standards. Consequently it can be recorded on any spare analog audio track on any type of recorder and distributed on any analog audio signal feed. This is exactly what is done on analog multi-track machines; one audio track is dedicated to timecode. In production VTRs a dedicated linear track may be provided. VTRs may also use vertical interval timecode (VITC). This is a form of timecode which is recorded in the vertical interval of the video signal so that it does not appear in the picture. The advantage of VITC is that it is still read out even when the video tape is not moving.

The timecode waveform is provided by a timecode generator. Such a device may be free standing or incorporated in a recorder. It may run free from a starting code set by the user, or it may lock to, or copy, a timecode signal sent from another generator elsewhere. Figure 4.35 shows an example of a VTR with a built-in timecode generator which records that timecode on its own tape as well as sending it to an analog audio recorder. The two tapes will then record the same timecode.

In order to play back the two tapes in lock, a timecode synchronizer is necessary. Traditionally this was a free-standing unit, but increasingly the function is built into the recorder. Figure 4.36 shows that a synchronizer requires two time-code readers, one for each tape. One of the recorders plays at constant speed; this is called the master. The timecode from the two machines is compared and used to vary the speed of the other; this is called the slave. In the case of an audio recorder and a video recorder the VTR would be the master because it is locked to station reference and cannot readily change speed. Once the two timecodes become the same, the two machines run at constant speed and in lip-sync.

As the timecode signal has a relatively low frequency, it can still be read at high speed. Consequently with suitable tape transports timecode can still be used to synchronize two tapes even if they are shuttled. This is known as chasing. Generally one machine will rewind faster than the other and will have to wait until the slower machine catches up before entering play mode again.

4.16 Digital audio disk systems

The disk drive was outlined in Section 4.10. In order to use disk drives for the storage of audio, a system like the one shown in Figure 4.37 is needed. The control computer determines where and when samples will be stored and retrieved, and sends instructions to the disk controller which causes the drives to read or write, and transfers samples between them and the memory. The instantaneous data rate of a typical drive is roughly ten times higher than the sampling rate, and this may result in the system data bus becoming choked by the disk transfers so other transactions are locked out. This is avoided by giving the disk controller DMA system a lower priority for bus access so that other devices can use the bus. A rapidly spinning disk cannot wait, and in order to prevent data loss, a silo or FIFO memory is necessary in the disk controller [4.13]. A silo will be interposed in the disk controller data stream in the fashion shown in Figure 4.38 so that it can buffer data both to and from the disk. When reading the disk, the silo starts to empty, and if there is bus contention, the silo will start to fill as shown in Figure 4.39(a). Where the bus is free, the disk controller will attempt to empty the silo into the memory. The system can take advantage of the interblock gaps

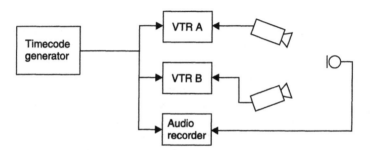

Figure 4.35 Using a single timecode generator, the same timecode recording can be put on more than one tape so that they can subsequently be played back synchronously.

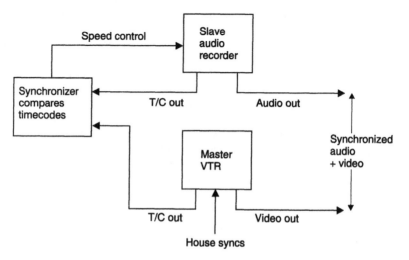

Figure 4.36 In a time code synchronizer system one of the machines is the master and the two timecode signals are compared to control the slave until they read the same.

on the disk, containing headers, preambles and redundancy, for in these areas there are no data to transfer, and there is some breathing space to empty the silo before the next block. In practice the silo need not be empty at the start of every block, provided it never becomes full before the end of the transfer. If this happens some data are lost and the function must be aborted. The block containing the silo overflow will generally be re-read on the next revolution. In sophisticated systems, the silo has a kind of dipstick which indicates how full it is, and can interrupt the central processing unit (CPU) if the data get too deep. The CPU can then suspend some bus activity to allow the disk controller more time to empty the silo.

When the disk is to be written, a continuous data stream must be provided during each block, as the disk cannot stop. The silo will be prefilled before the disk attempts to write as shown in Figure 4.39(b), and the disk controller attempts to keep it full. In this case all will be well if the silo does not become empty before the end of the transfer.

The disk controller cannot supply samples at a constant rate, because of gaps between blocks, defective blocks and the need to move the heads from one track to another and because of system bus contention. In order to accept a steady audio sample stream for storage, and to return it in the same way on replay, hard disk based audio recorders must have a quantity of RAM for buffering. Then there is time for the positioner to move whilst the audio output is supplied from the RAM. In replay, the drive controller attempts to keep the RAM as full as possible by issuing a read command as soon as one block space appears in the

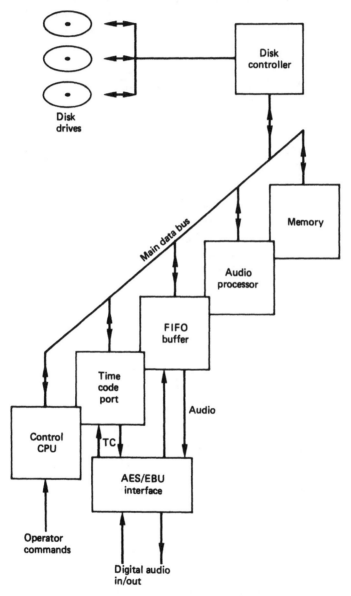

Figure 4.37 The main parts of a digital audio disk system. Memory and FIFO allow continuous audio despite the movement of disk heads between blocks.

RAM. This allows the maximum time for a seek to take place before reading must resume. Figure 4.40 shows the action of the RAM during reading. Whilst recording, the drive controller attempts to keep the RAM as empty as possible by issuing write commands as soon as a block of data is present, as in Figure 4.41. In this way the amount of time available to seek is maximized in the presence of a continuous audio sample input.

A hard disk has a discontinuous recording and acts like a RAM, in that it must be addressed before data can be retrieved. The rotational speed of the disk is constant and not locked to anything. A vital step in converting a disk drive into an audio recorder is to establish a link between the time through the recording and

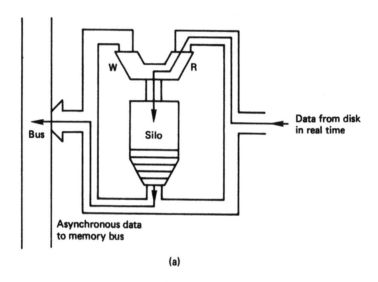

(a)

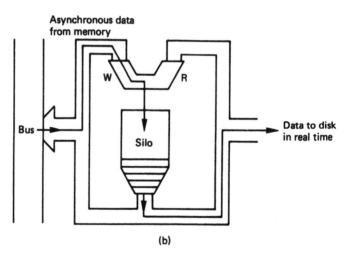

(b)

Figure 4.38 In order to guarantee that the drive can transfer data in real time at regular intervals (determined by disk speed and density) the silo provides buffering to the asynchronous operation of the memory access process. In (a) the silo is configured for a disk read. The same silo is used in (b) for a disk write.

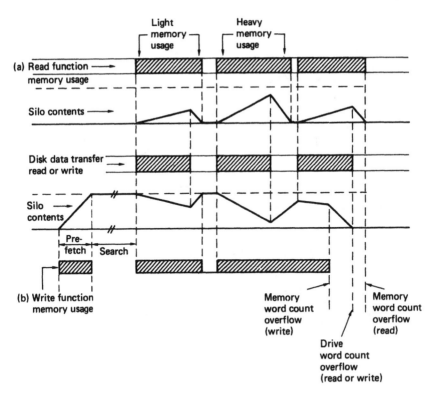

Figure 4.39 The silo contents during read functions (a) appear different from those during write functions (b). In (a), the control logic attempts to keep the silo as empty as possible; in (b) the logic prefills the silo and attempts to keep it full until the memory word count overflows.

the location of the data on the disk.

When audio samples are fed into a disk-based system, from an AES/EBU interface or from a converter, they will be placed initially in RAM, from which the disk controller will read them by DMA. The continuous-input sample stream will be split up into disk blocks for disk storage. Timecode will be used to assemble a table which contains a conversion from real time in the recording to the physical disk address of the corresponding audio files. In a sound for television application, the timecode would be the same as that recorded on the videotape. Wherever possible, the disk controller will allocate incoming audio samples to contiguous disk addresses, since this eases the conversion from timecode to physical address [4.14]. This is not, however, always possible in the presence of defective blocks, or if the disk has become chequerboarded from repeated re-recording.

The table of disk addresses will also be made into a named disk file and stored in an index which will be in a different area of the disk from the audio files. Several recordings may be fed into the system in this way, and each will have an entry in the index.

If it is desired to play back one or more of the recordings, then it is only necessary to specify the starting timecode and the filename. The system will look

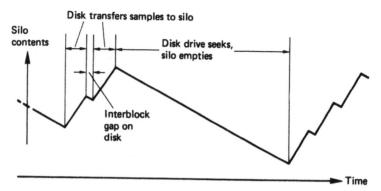

Figure 4.40 During an audio replay sequence, silo is constantly emptied to provide samples, and is refilled in blocks by the drive.

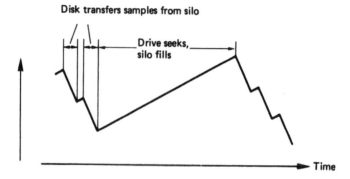

Figure 4.41 During audio recording, the input samples constantly fill the silo, and the drive attempts to keep it empty by reading from it.

up the index file in order to locate the physical address of the first and subsequent sample blocks in the desired recording, and will begin to read them from disk and write them into the RAM. Once the RAM is full, the real time replay can begin by sending samples from RAM to the output or to local converters. The sampling rate clock increments the RAM address and the timecode counter. Whenever a new timecode frame is reached, the corresponding disk address can be obtained from the index table, and the disk drive will read a block in order to keep the RAM topped up.

The disk transfers must by definition take varying times to complete because of the rotational latency of the disk. Once all the sectors on a particular cylinder have been read, it will be necessary to seek to the next cylinder, which will cause a further extension of the reading sequence. If a bad block is encountered, the sequence will be interrupted until it has passed. The RAM buffering is sufficient to absorb all of these access time variations. Thus the RAM acts as a delay between the disk transfers and the sound which is heard. A corresponding advance is arranged in timecodes fed to the disk controller. In effect the actual timecode has a constant added to it so that the disk is told to obtain blocks of samples in advance of real time. The disk takes a varying time to obtain the samples, and the RAM then further delays them to the correct timing. Effectively the disk/RAM subsystem is a timecode controlled memory. One need only put in the time, and out comes the audio corresponding to that time. This is the characteristic of an audio synchronizer. In most audio equipment the synchronizer is extra; the hard disk needs one to work at all, and so every hard disk comes with a free synchronizer. This makes disk-based systems very flexible as they can be made to lock to almost any reference and care little what sampling rate is used or if it varies. They perform well locked to videotape or film via timecode because no matter how the pictures are shuttled or edited, the timecode link always produces the correct sound to go with the pictures. A video edit decision list (EDL) can be used to provide an almost immediate rough edited soundtrack.

A multi-track recording can be stored on a single disk and, for replay, the drive will access the files for each track faster than real time so that they all become present in the memory simultaneously. It is not, however, compulsory to play back the tracks in their original time relationship. For the purpose of synchronization [4.15] or other effects, the tracks can be played with any time relationship desired, a feature not possible with multi-track tape drives.

4.17 Audio in VTRs

In most analog video recorders, the audio tracks are longitudinal and are placed at one edge of the tape. Figure 4.42 shows a typical analog tape track arrangement. The audio performance of video recorders has traditionally lagged behind that of audio-only recorders. In video recorders, the use of rotary heads to obtain sufficient bandwidth results in a wide tape which travels relatively slowly by professional audio standards. The audio performance is limited by the format itself and by the nature of video recording. In all rotary-head recorders, the intermittent head contact causes shock-wave patterns to propagate down the tape, making low flutter figures difficult to achieve. This is compounded by the action of the capstan servo which has to change tape speed to maintain control-track phase if the video heads are to track properly.

The requirements of lip-sync dictate that the same head must be used for both recording and playback, when the optimum head design for these two functions is different. When dubbing from one track to the next, one head gap will be recording the signal played back by the adjacent magnetic circuit in the head, and

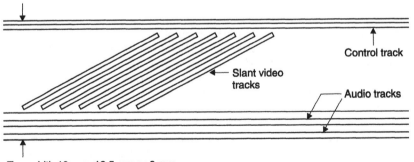

Figure 4.42 Track layout of an analog VTR. Note the linear audio tracks are relatively narrow. The linear tape speed is quite low by audio recording standards.

mutual inductance can cause an oscillatory loop if extensive antiphase crosstalk cancelling is not employed. Placing the tracks on opposite sides of the tape would help this problem, but gross phase errors between the channels would then be introduced by tape weave. This can mean the difference between a two-channel recorder and a stereo recorder. Even when the audio channels are on the same edge of the tape, most analog video recorders have marginal interchannel phase accuracy. As the linear tape speed is low, recorded wavelengths are short and a given amount of tape weave results in greater phase error. Crosstalk between the timecode and audio tracks can also restrict performance. The introduction of stereo audio with television and the modern trend toward extensive post production makes sound quality and the ability to support multi-generation work essential.

In consumer recorders, economy of tape consumption is paramount. It is also important in portable professional recorders used for electronic news gathering (ENG). In videocassette recorders such as VHS and Betamax, and in the professional Betacam and M-II formats, the rotary head combined with azimuth recording allowed extremely narrow tape tracks, with the result that linear tape speed was very low. The analog audio tracks of these systems gave marginal performance.

One solution to the audio quality issue was to frequency-modulate the audio onto a carrier which was incorporated in the spectrum of the signals recorded by the rotary heads. The audio quality of such systems was very much better than that offered by the analog tracks, but there was a tremendous drawback for production applications in that the audio could not be recorded independently of the video, or vice versa, as they were both combined into one signal.

The adoption of digital techniques essentially removes these problems for the audio in VTRs just as it does for audio-only recorders. Once the audio is in numerical form, wow, flutter and channel-phase errors can be eliminated by timebase correction; crosstalk ceases to occur and, provided a suitable error correction strategy is employed, the only degradation of the signal will be due to quantizing. The most significant advantages of digital recording are that there is essentially no restriction on the number of generations of re-recording which can be used and that proper crossfades can be made in the audio at edit points, following a rehearsal if necessary.

4.18 Audio in digital VTRs

Digital video recorders are capable of exceptional picture quality, but this attribute seems increasingly academic. The real power of digital technology in broadcasting is economic. DVTRs cost less to run than analog equipment. They use less tape and need less maintenance. Cassette formats are suitable for robotic handling, reducing manpower requirements.

A production VTR must offer such features as still frame, slow motion, timecode, editing, pictures in shuttle and so on, as well as the further features needed in the digital audio tracks, such as the ability to record the user data in the AES/EBU digital audio interconnect standard. It must be possible to edit the video and each of the audio channels independently, whilst maintaining lip-sync.

An outline of a representative DVTR follows to illustrate some general principles, and to put the audio subsystem in perspective. Figure 4.43 shows that a DVTR generally has converters for both video and audio. Once in digital form, the data are formed into blocks, and the encoding section of the error-correction system supplies redundancy designed to protect the data against errors. These blocks are then converted into some form of channel code which combines the data with clock information so that it is possible to identify how many bits were recorded even if several adjacent bits are identical. The coded data are recorded on tape by a rotary head. Upon replaying the recording of the hypothetical machine of Figure 4.43, the errors caused by various mechanisms will be detected, corrected or concealed using the extra bits appended during the encoding process.

Digital tape tracks are always narrower than analog tracks, since they carry binary data, and a large signal-to-noise ratio is not required. This helps to balance the greater bandwidth required by a digitized video signal, but it does put extra demands on the mechanical accuracy of the transport in order to register the heads with the tracks. The prodigious data rate requires high frequency circuitry, which causes headaches when complex processing such as error correction is contemplated. Unless a high compression factor is used, DVTRs often use more than one data channel in parallel to reduce the data rate in each. Figure 4.43 shows the distribution of incoming audio and video data between two channels. At any one time there are two heads working on two parallel tape tracks in order to share the data rate.

In the digital domain, samples can be recorded in any structure whatsoever provided they are output in the correct order on replay. This means that a segmented scan is acceptable without performance penalty in a digital video recorder. A segmented scan breaks up each television picture into a number of different sweeps of the head across the tape. This permits a less than complete wrap of the scanner, which eases the design of the cassette-threading mechanism. The use of very short recorded wavelengths in digital recording makes the effect of spacing loss due to contamination worse, and so a cassette system is an ideal way of keeping the medium clean.

The use of several cassette sizes satisfies the contradicting demands of a portable machine which must be small and a studio machine which must give good playing time. The hubs of DVTR cassettes are at different spacings in the different sizes, so a machine built to play more than one size will need moving reel-motors. The cassette shell contains patterns of holes which can be read by sensors in the player. One set of these can be used to allow tapes of different thickness to change the tape remaining calculation based on the hub speed automatically. The other set is intended for the user, and one of these causes total record-lockout, or write-protect, when a plug is removed. Another may allow only the audio to be recorded, locking out the recording of the video and control tracks.

The audio samples in a DVTR are binary numbers just like the video samples, and although there is an obvious difference in sampling rate and wordlength, this only affects the relative areas of tape devoted to the audio and video samples. The most important difference between audio and video samples is the tolerance to errors. The acuity of the ear means that uncorrected audio samples must not occur more than once every few hours. There is little redundancy in sound, and concealment of errors is not desirable on a routine basis. In video, the samples

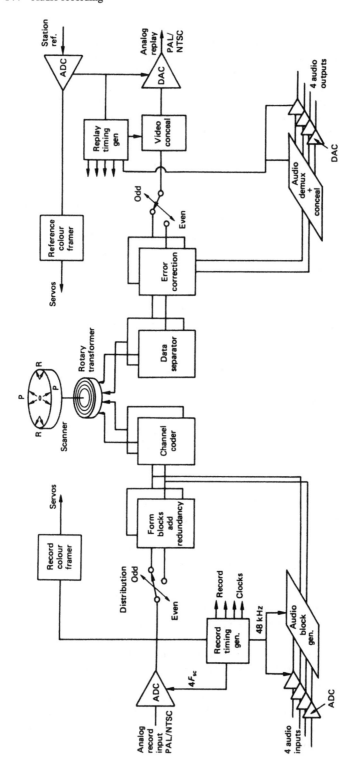

Figure 4.43 Analog video is sampled at 4F$_{SC}$ and audio at 48 kHz. Samples are distributed between two channels, formed into blocks, and coded for recording. The audio blocks are time shared with video through a common tape channel consisting of a channel coder, the rotating head assembly and the data separator. On replay, timing is generated by station reference. Errors which cannot be corrected are concealed before conversion.

are highly redundant, and concealment can be effected using samples from previous or subsequent lines or, with care, from the previous frame. Thus there are major differences between the ways that audio and video samples are handled in a DVTR. One such difference is that the audio samples often have 100% redundancy: every one is recorded using about twice as much space on tape as the same amount of video data.

As digital audio and video are both just data, in DVTRs the audio samples are carried by the same channel as the video samples. This reduces the number of encoders, preamplifiers and data separators needed in the system, whilst increasing the bandwidth requirement by only a few percent even with double recording. Figure 4.44 shows that the audio samples are heavily time compressed so that the audio and video data have the same bit rate on tape. This increases the amount of circuitry which can be common to both video and audio processing.

In order to permit independent audio and video editing, the tape tracks are given a block structure. Editing will require the heads momentarily to go into record as the appropriate audio block is reached. Accurate synchronization is necessary if the other parts of the recording are to remain uncorrupted. The concept of a head which momentarily records in the centre of a track which it is reading is the normal operating procedure for all computer disk drives. There are in fact many parallels between digital helical recorders and disk drives, including the use of the term *sector*. In moving-head disk drives, the sector address is a measure of the angle through which the disk has rotated. This translates to the phase of the scanner in a rotary-head machine. The part of a track which is in one sector is called a block. The word 'sector' is often used instead of 'block' in casual parlance when it is clear that only one head is involved. However, as many DVTRs have two heads in action at any one time, the word 'sector' means the two side-by-side blocks in the segment. As there are four independently recordable audio channels in most production DVTR formats, there are four audio sectors.

There is a requirement for the DVTR to produce pictures in shuttle. In this case, the heads cross tracks randomly, and it is most unlikely that complete video blocks can be recovered. To provide pictures in shuttle, each block is broken down into smaller components called sync blocks in the same way as is done in DAT. These contain their own error checking and an address, which in disk terminology would be called a header, which specifies where in the picture the samples in the sync block belong. In shuttle, if a sync block is read properly, the address can be used to update a frame store. Thus it can be said that a sector is the smallest amount of data which can be written and is that part of a track pair within the same sector address, whereas a sync block is the smallest amount of data which can be read. Clearly there are many sync blocks in a sector.

The same sync block structure continues in the audio because the same read/write circuitry is used for audio and video data. Clearly the address structure must also continue through the audio. In order to prevent audio samples from arriving in the frame store in shuttle, the audio addresses are different from the video addresses. In all formats, the arrangement of the audio blocks is designed to maximize data integrity in the presence of tape defects and head clogs. The allocation of the audio channels to the sectors may change from on segment to the next. If a linear tape scratch damages the data in a given audio channel in one segment, it will damage a different audio channel in the next. Thus the scratch damage is shared between all four audio channels, each of which need correct

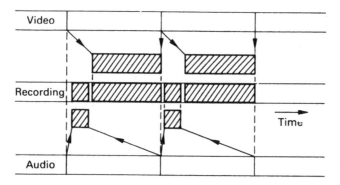

Figure 4.44 Time compression is used to shorten the length of track needed by the video. Heavily time-compressed audio samples can then be recorded on the same track using common circuitry

only one-quarter of the damage. The relationship of the audio channels to the physical tracks may rotate by one track against the direction of tape movement from one audio sector to the next. If a head becomes clogged, the errors will be distributed through all audio channels, instead of causing severe damage in one channel.

Clearly the number of audio sync blocks in a given time is determined by the number of video fields in that time. It is only possible to have a fixed tape structure if the audio sampling rate is locked to video. With 625/50 machines, the sampling rate of 48 kHz results in exactly 960 audio samples in every field.

For use on 525/60, it must be recalled that the 60 Hz is actually 59.94 Hz. As this is slightly slow, it will be found that in 60 fields, exactly 48 048 audio samples will be necessary. Unfortunately 60 will not divide into 48 048 without a remainder. The largest number which will divide 60 and 48 048 is 12; thus in $60/12 = 5$ fields there will be $48\,048/12 = 4004$ samples. Over a five-field sequence the product blocks contain 801, 801, 801, 801 and 800 samples respectively, adding up to 4004 samples.

In order to comply with the AES/EBU digital audio interconnect, wordlengths between 16 and 20 bits can be supported by most DVTRs, but it is necessary to record a code in the sync block to specify the wordlength in use. Pre-emphasis may have been used prior to conversion, and this status is also to be conveyed, along with the four channel-use bits. The AES/EBU digital interface (see Chapter 3) uses a block-sync pattern which repeats after 192 sample periods corresponding to 4 ms at 48 kHz. Since the block size is different to that of the DVTR interleave block, there can be any phase relationship between interleave-block boundaries and the AES/EBU block-sync pattern. In order to re-create the same phase relationship between block sync and sample data on replay, it is necessary to record the position of block sync within the interleave block. It is the function of the interface control word in the audio data to convey these parameters. There is no guarantee that the 192-sample block-sync sequence will remain intact after audio editing; most likely there will be an arbitrary jump in block-sync phase. Strictly speaking a DVTR playing back an edited tape would have to ignore the block-sync positions on the tape, and create new block sync. at the standard 192-sample spacing. Unfortunately most DVTR formats are not totally transparent to the whole of the AES/EBU data stream.

4.19 Editing

In all types of audio editing the goal is the appropriate sequence of sounds at the appropriate times. Figure 4.45 shows how a master recording is assembled from source recordings. In analog audio equipment, editing was almost always performed using tape or magnetically striped film. These media have the characteristic that the time through the recording is proportional to the distance along the track. Editing consisted of physically cutting and splicing the medium, in order to assemble the finished work mechanically, or of copying lengths of source medium to the master.

When this was the only way of editing, it did not need a qualifying name. Now that audio is stored as data, alternative storage media have become available which allow editors to reach the same goal but using different techniques. Whilst open-reel digital audio tape formats support splice editing, in all other digital audio editing samples from various sources are brought from the storage media to various pages of RAM. The edit is performed by crossfading between sample streams retrieved from RAM and subsequently rewriting on the output medium. Thus the nature of the storage medium does not affect the form of the edit in any way except the amount of time needed to execute it.

Tapes only allow serial access to data, whereas disks and RAM allow random access and so can be much faster. Editing using random access storage devices is very powerful as the shuttling of tape reels is avoided. The technique is sometimes called non-linear editing. This is not a very helpful name, as in these systems the editing itself is performed in RAM in the same way as before. In fact it is only the time axis of the storage medium which is non-linear.

Digital recording media vary in their principle of operation, but all have in common the use of error correction. This requires an interleave, or reordering, of samples to reduce the impact of large errors, and the assembling of many samples into an error-correcting code word. Code words are recorded in constant sized blocks on the medium. Audio editing requires the modification of source material in the correct real time sequence to sample accuracy. This contradicts the interleaved block based codes of real media.

Editing to sample accuracy simply cannot be performed directly on real media. Even if an individual sample could be located in a block, replacing the samples after it would destroy the codeword structure and render the block uncorrectable.

The only solution is to ensure that the medium itself is only edited at block boundaries so that entire error correction codewords are written down. In order to obtain greater editing accuracy, blocks must be read from the medium and de-interleaved into RAM, modified there and re-interleaved for writing back on the medium, the so-called *read-modify-write* process.

In disks, blocks are often associated into clusters which consist of a fixed number of blocks in order to increase data throughput. When clustering is used, editing on the disk can only take place by rewriting entire clusters.

The digital audio editor consists of three main areas. Firstly, the various contributory recordings must enter the processing stage at the right time with respect to the master recording. This will be achieved using a combination of timecode, transport synchronization and RAM timebase correction. The synchronizer will take control of the various transports during an edit so that one section reaches its

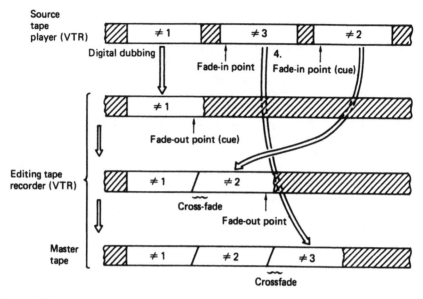

Figure 4.45 The function of an editor is to perform a series of assembles to produce a master tape from source tapes.

out-point just as another reaches its in-point.

Secondly the audio signal path of the editor must take the appropriate action, such as a crossfade, at the edit point. This requires some digital processing circuitry.

Thirdly the editing operation must be supervised by a control system which co-ordinates the operation of the transports and the signal processing to achieve the desired result.

Figure 4.46 shows a simple block diagram of an editor. Each source device, be it disk or tape or some other medium, must produce timecode locked to the audio samples. The synchronizer section of the control system uses the timecode to determine the relative timing of sources and sends remote control signals to the transport to make the timing correct. The master recorder is also fed with time-code in such a way that it can make a contiguous timecode track when perform-ing assembly edits. The control system also generates a master sampling rate clock to which contributing devices must lock in order to feed samples into the edit process. The audio signal processor takes contributing sources and mixes them as instructed by the control system. The mix is then routed to the recorder.

Digital audio editors must simulate the 'rock and roll' process of edit-point location in analog tape recorders where the tape reels are moved to and fro by hand. Most digital tape recorders and CD players can only play at or close to nor-mal speed. Disk drives transfer data intermittently much faster than real time. The solution is to transfer the recording in the area of the edit point to RAM in the editor. RAM access can take place at any speed or direction and the precise edit point can then be conveniently found by monitoring audio from the RAM.

Figure 4.47 shows how the area of the edit point is transferred to the memory. The source device is commanded to play, and the operator listens to replay samples via a DAC in the monitoring system. The same samples are continuously written into a memory within the editor. This memory is addressed by a counter which repeatedly overflows to give the memory a ring-like structure rather like that of a timebase corrector, but somewhat larger. When the operator hears the rough area in which the edit is required, he will press a button. This action stops the memory writing, not immediately, but one-half of the memory contents later. The effect is then that the memory contains an equal number of samples before and after the rough edit point.

Once the recording is in the memory, it can be accessed at leisure, and the constraints of the source device play no further part in the edit-point location. There are a number of ways in which the memory can be read. If the memory address is supplied by a counter which is clocked at the appropriate rate, the edit area can be replayed at normal speed, or at some fraction of normal speed repeatedly. In order to simulate the analog method of finding an edit point, the operator is provided with a *scrub wheel* or rotor, and the memory address will change at a rate proportional to the speed with which the rotor is turned, and in the same direction. Thus the sound can be heard forward or backward at any speed, and the effect is exactly that of manually rocking an analog tape past the heads of an ATR.

Samples which will be used to make the master recording need never pass through these processes; they are solely to assist in the location of the edit points. The sound quality in this mode can be impaired to various degrees by the sampling rate converter and any data reduction used, but this does not affect the finished work.

The act of pressing the coarse edit-point button stores the timecode of the source at that point, which is frame-accurate. As the rotor is turned, the memory address is monitored, and used to update the timecode to sample accuracy.

Before assembly can be performed, two edit points must be determined, the out-point at the end of the previously recorded signal, and the in-point at the beginning of the new signal. The editor's microprocessor stores these in an edit decision list (EDL) in order to control the automatic assemble process.

4.20 Non-linear editing

Using one or other of the above methods, an edit list can be made which contains an in-point, an out-point and an audio filename for each of the segments of audio which need to be assembled to make the final work, along with a crossfade period and a gain parameter. This edit list will also be stored on the disk. When a preview of the edited work is required, the edit list is used to determine what files will be necessary and when, and this information drives the disk controller.

Figure 4.48 shows the events during an edit between two files. The edit list causes the relevant audio blocks from the first file to be transferred from disk to memory, and these will be read by the signal processor to produce the preview output. As the edit point approaches, the disk controller will also place blocks from the incoming file into the memory. It can do this because the rapid data-transfer rate of the drive allows blocks to be transferred to memory much faster than real time, leaving time for the positioner to seek from one file to another. In different areas of the memory there will be simultaneously the end of

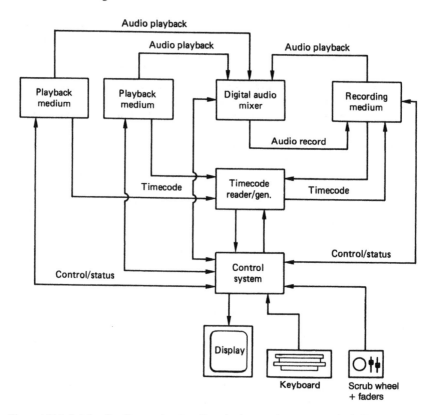

Figure 4.46 A digital audio editor requires an audio path to process the samples, and a timing and synchronizing section to control the time alignment of signals from the various sources. A supervisory control system acts as the interface between the operator and the hardware.

the outgoing recording and the beginning of the incoming recording. The signal processor will use the fine edit-point parameters to work out the relationship between the actual edit points and the cluster boundaries. The relationship between the cluster on disk and the RAM address to which it was transferred is known, and this allows the memory address to be computed in order to obtain samples with the correct timing. Before the edit point, only samples from the outgoing recording, are accessed, but as the crossfade begins, samples from the incoming recording are also accessed, multiplied by the gain parameter and then mixed with samples from the outgoing recording, according to the crossfade period required. The output of the signal processor becomes the edited preview material, which can be checked for the required subjective effect. If necessary the in-or out-points can be trimmed, or the crossfade period changed, simply by modifying the edit-list file. The preview can be repeated as often as needed, until the desired effect is obtained. At this stage the edited work does not exist as a file, but is re-created each time by a further execution of the EDL. Thus a lengthy editing session need not fill up the disk.

It is important to realize that at no time during the edit process were the original audio files modified in any way. The editing was done solely by reading the audio files. The power of this approach is that if an edit list is created

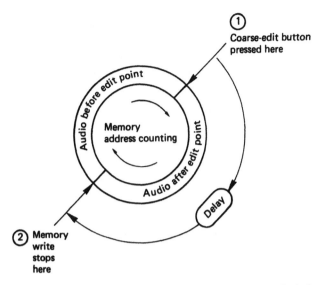

Figure 4.47 The use of a ring memory which overwrites allows storage of samples before and after the coarse-edit point.

wrongly, the original recording is not damaged, and the problem can be put right simply by correcting the edit list. The advantage of a disk-based system for such work is that location of edit points, previews and reviews are all performed almost instantaneously, because of the random access of the disk. This can reduce the time taken to edit a programme to a quarter of that needed with a tape machine [4.16].

During an edit, the disk drive has to provide audio files from two different places on the disk simultaneously, and so it has to work much harder than for a simple playback. If there are many close-spaced edits, the drive may be hard-pressed to keep ahead of real time, especially if there are long crossfades, because during a crossfade the source data rate is twice as great as during replay. A large buffer memory helps this situation because the drive can fill the memory with files before the edit actually begins, and thus the instantaneous sample rate can be met by the memory's emptying during disk-intensive periods. In practice, crossfades measured in seconds can be achieved in a disk-based system, a figure which is not matched by tape systems.

Disk formats which handle defects dynamically, such as defect skipping, will also be superior to bad-block files when throughput is important. Some drives rotate the sector addressing from one cylinder to the next so that the drive does not lose a revolution when it moves to the next cylinder. Disk-editor performance is usually specified in terms of peak editing activity which can be achieved, but with a recovery period between edits. If an unusually severe editing task is necessary where the drive just cannot access files fast enough, it will be necessary to rearrange the files on the disk surface so that files which will be needed at the same time are on nearby cylinders [4.17]. An alternative is to spread the material between two or more drives so that overlapped seeks are possible.

Once the editing is finished, it will generally be necessary to transfer the edited material to form a contiguous recording so that the source files can make

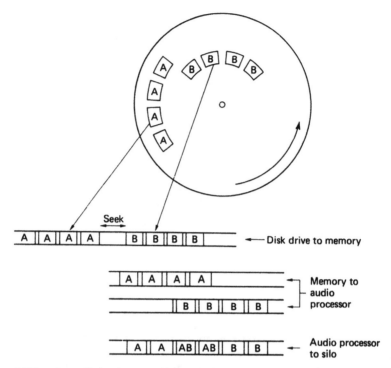

Figure 4.48 In order to edit together two audio files, they are brought to memory sequentially. The audio processor accesses file pages from both together, and performs a crossfade between them. The silo produces the final output at constant steady-sampling rate.

way for new work. If the source files already exist on tape the disk files can simply be erased. If the disks hold original recordings they will need to be backed up to tape if they will be required again. In large broadcast systems, the edited work can be broadcast directly from the disk file. In smaller systems it will be necessary to output to some removable medium, since the Winchester drives in the editor have fixed media. It is only necessary to connect the AES/EBU output of the signal processor to any type of digital recorder, and then the edit list is executed once more. The edit sequence will be performed again, exactly as it was during the last preview, and the results will be recorded on the external device.

References

4.1   Mallinson, J.C. *The Foundations of Magnetic Recording*, Academic Press (1987)
4.2   Mee, C.D. *The Physics of Magnetic Recording*, Amsterdam: Elsevier–North Holland Publishing (1978)
4.3   Mee, C.D. and Daniel, E.D. (eds). *Magnetic Recording*, Vol.III, Ch. 3: McGraw-Hill (1988)
4.4   Bertram, H.N. Long wavelength AC bias recording theory. *IEEE Trans. Magn.* **MAG 10,** 1039 (1974)

4.5 Dolby, R.M., An audio noise reduction system, *J. Audio Eng. Soc.*, **15,** 383 (1967)

4.6 Watkinson, J.R. *The Art of Data Recording*, Ch. 3. Oxford: Focal Press (1994)

4.7 Deeley, E.M. Integrating and differentiating channels in digital tape recording. *Radio Electron. Eng.*, **56** 169–173 (1986)

4.8 Nakajima, H. and Odaka, K. A rotary-head high-density digital audio tape recorder. *IEEE Trans. Consum. Electron.*, **CE-29**, 430–437 (1983)

4.9 Watkinson, J.R. *The Art of Digital Audio*, Ch. 6. Oxford: Focal Press (1994)

4.10 Watkinson, J.R. *The Art of Data Recording*, Ch. 4. Oxford: Focal Press (1994)

4.11 Fukuda, S., Kojima, Y., Shimpuku, Y. and Odaka, K. 8/10 modulation codes for digital magnetic recording. *IEEE Trans. Magn.*, **MAG-22**, 1194–1196 (1986)

4.12 Ratcliff, J. *Time Code*, Oxford: Focal Press (1993)

4.13 Ingbretsen, R.B. and Stockham, T.G. Random access editing of digital audio. *J. Audio Eng. Soc.*, **32**, 114–122 (1982)

4.14 McNally, G.W., Gaskell, P.S. and Stirling, A.J. Digital audio editing. *BBC Research Dept Report*, RD 1985/10 (1985)

4.15 McNally, G.W., Bloom, P.J. and Rose, N.J. A digital signal processing system for automatic dialogue post-synchronisation. 82nd Audio Eng. Soc. Convention, preprint 2476(K-6). London: Audio Eng. Soc. (1987)

4.16 Todoroki, S. *et al.*, New PCM editing system and configuration of total professional digital audio system in near future. 80th Audio Eng. Soc. Convention, preprint 2319(A8). Montreux: Audio Eng Soc (1986)

4.17 McNally, G.W., Gaskell, P.S. and Stirling, A.J. Digital audio editing. *BBC Research Dept. Report*, RD 1985/10 (1985)

Chapter 5

Audio routing and transmission

5.1 Introduction

Audio signals need to be transmitted from one place to another both over short distances to a single destination during the production process and over long distances to many destinations during broadcasting. In television the audio and video signals will frequently be combined in some way. In this chapter the principles of audio routing and broadcasting in the context of television will be treated along with the important means of synchronization needed.

Typical installations generally contain a combination of analog and digital audio equipment linked by routers and converters. The configuration may be redundant to allow continued use in the case of equipment failures, or expensive, special-purpose devices may be shared between studios or edit suites. It will be seen here that the reconfiguration made possible by the use of routers places heavy demands on signal standardization. It does not matter whether the system is analog or digital; correct operation can only be ensured if all devices in a signal chain are configured so that the meaning of the signal does not change with the configuration. Analog signals have standardized voltages and timings; digital signals have standardized code values and protocols.

To ensure that standards are met, various tests have been developed. These are described here, along with the equipment and techniques needed to make them. The testing procedures for analog and digital equipment are quite different. However, in order correctly to set up an ADC or a DAC, both types of test may be needed together.

5.2 Routers

In a typical audio production system the signal may pass through a large number of different devices on its way from the microphone to the final viewer. For practical reasons, not least the ability to carry on working if an individual item fails, most installations connect major signal processing blocks to a central switching device called a *router*, which is essentially a telephone exchange for video, audio and control signals. Figure 5.1(a) shows a system without a router. As every device is effectively in a chain, the failure of one device breaks the chain and the whole system fails. The alternative is the system of Figure 5.1(b) in which the same equipment is connected to a router. It will be seen that the router functions as a *crosspoint switch* – a device capable of connecting any vertical line to any horizontal line. The crosspoints in Figure 5.1(b) may be set to give exactly the same signal flow as in (a). However, in the case of a failure, a rearrangement of the crosspoints will allow the signal through, bypassing the failed device. Obviously the router itself is expected to be extremely reliable. Generally the electronic switching circuits are fed from a dual power supply, each half of which is capable of delivering the entire power requirements. In the case of a power supply fault there is no interruption to the signal. It is normal practice to connect each power supply to a different phase of the electricity supply so that a single-phase failure cannot take out the router.

Whilst a breakdown in a post production process causes delay, a breakdown in the on-air signal path will be highly unpopular and so systems are designed to be fault tolerant. Any equipment designed for use on-air must not only be designed for reliability, but also with the assumption that it will go wrong at some time. Figure 5.2 shows a device which contains a *bypass relay*. In the case of power loss or an internally detected fault condition, the bypass relay will automatically connect the input direct to the output. The process performed by the unit will not now be available, but the signal path is not broken, except momentarily. In practice this is not as bad as it seems because in practical installations, many signal processing stages are present in case an incoming signal is deficient in some way so that a correction can be made. Most of the time, many of these units do not alter the signal at all and so their omission from the signal

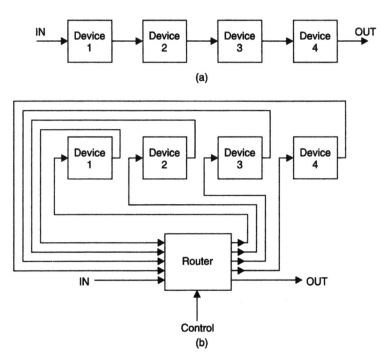

Figure 5.1 (a) Without a router, the system configuration can only be changed by rearranging interconnecting wiring. If a device fails, it cannot readily be bypassed. (b) With a router, the system configuration is flexible. Devices can be omitted from the chain for maintenance.

path is of little consequence. The bypass relay allows the unit to pass a signal in the case of failure, but it does not allow the unit to be removed for maintenance. In this case the router would be re-configured to bypass the device and its wiring completely.

In large installations routers can be used to make the system more fault tolerant or more economical according to requirements. The two goals are usually exclusive and so the philosophy of the system must be clear at the design stage. In a critical installation, such as a presentation suite where signals are actually connected live to the transmitter, multiple redundancy may be employed. Figure 5.3(a) shows a television station which transmits two channels. Here the inputs of three routers are connected in parallel to all audio source signals. Three audio mixers are available, two of which will be on-air at any one time. A second, smaller, router determines which pair of mixers are on-air. Any one audio mixer or router can fail but the system will not be permanently affected.

Figure 5.3(b) shows a pair of edit suites which do not work on air and can therefore tolerate failures. For economy, a single artificial reverberator is connected to the router. It can process signals from one suite or the other, but not both. A single device shared between systems in this way is said to be *assigned* to the system it is connected to.

5.3 Multilayer routers

A router does not just switch audio signals. Generally it will also switch video and a number of control signals as well. Figure 5.4 shows a multilayer router having a common control system which is generally programmemed to switch every layer identically. In the case of a video recorder which can be assigned to two or more edit suites, the router will need to switch audio, video, timecode and the remote control signals which the edit controller uses to control the tape transport. In broadcast applications, one router layer may be for tally control. The tally signal starts at the feed to the transmitter as a DC voltage which travels in the opposite direction to the video signal. Figure 5.5 shows that at every stage where there is a routing decision, one layer of the router switches tally power

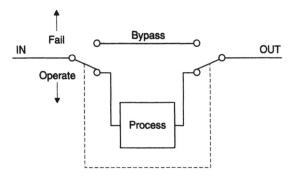

Figure 5.2 Using a bypass relay, a failing device can take itself out of service by connecting the input to the output.

towards the source of the selected signal. This process may be followed through several layers until the tally signal arrives at a signal source such as a microphone, a camera or a playout VTR. The DC power causes the 'on-air' light to illuminate so that all concerned know that unit is in the transmission path. If a different source is selected, the tally routing will automatically illuminate a different 'on-air' light.

As a simple router is no more than a number of switches, either relay contacts or electronic devices, then a reconfiguration of the router may result in an audible click. The disturbance will be greater if a digital audio signal is switched as the receiver may lose lock and the sound may mute briefly until synchronism is

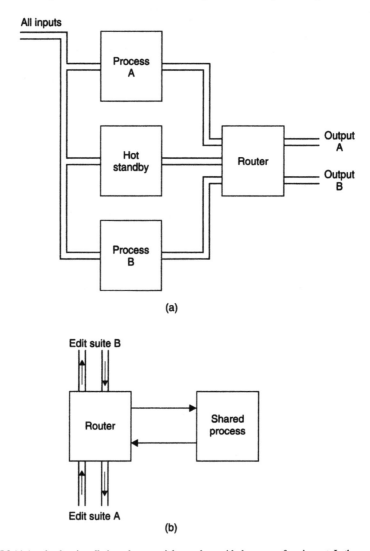

Figure 5.3 (a) A redundant installation where two jobs are done with three sets of equipment. In the case of a failure the hot spare is routed to the affected channel. (b) A shared installation where an expensive processing device can be assigned to whichever system needs it.

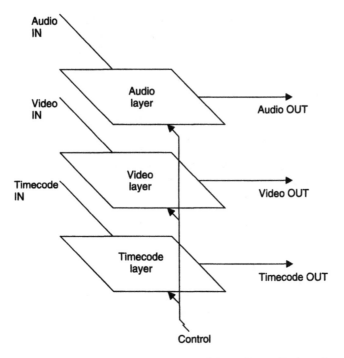

Figure 5.4 A multilayer router has switching for several kinds of signal: video, audio, timecode, etc., using a common control system.

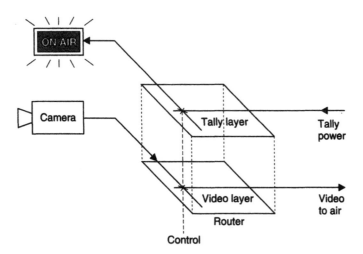

Figure 5.5 The tally system allows a control voltage to retrace the routing path taken by the video so that the 'on-air' lights reflect the camera or source which is currently live.

established. If the reconfiguration is simply to change the assignment of a signal processor, the click will be of no consequence. However if it is possible that switching will take place whilst on-air signals are passing through, so-called hot cutting, then clicks are unacceptable.

Figure 5.6 shows how a click-free hot cutting router can be made by assigning a simple two-input cross-fader to the router. When the fader is set completely to one input, the other input can be switched. The fader is then transitioned to the other input to give a click-free changeover. The first input can then be switched. In the analog domain the cross-fader will be implemented with VCAs (voltage-controlled amplifiers), whereas a digital cross-fader will use multipliers.

In analog routing, the router and associated cabling can affect the signal quality. The advantage of digital routing is that only data are being transmitted and so the transmission and routing process can be transparent. In a digital router the input audio signal is sliced and phase locked back to a clean square binary signal so that the routing process itself is carried out on logic levels. A clean digital signal is launched at the router output.

Digital audio routers can be implemented with other approaches than the simple crosspoint switch. In the time division multiplexed (TDM) router, the incoming audio signals are converted to the parallel domain and gated onto a common bus one after the other. This is done rapidly enough so that all input samples have been scanned in less than one sample period. Any bus receiver can obtain any channel simply by sampling the multiplexed bus at the appropriate time during the cycle. TDM systems are naturally synchronous, and when switching takes place, the structure of the AES/EBU signal will not be disturbed and there will be no muting, although clicks may result on a hot cut because of a discontinuity in the audio waveform.

It is, however, very easy to incorporate a hot cutting cross-fader into a TDM router. One bus receiver device simply samples two channels, cross-fades them and returns the mix to a later time slot with a gated bus driver. This blurs the distinction between audio mixing and routing.

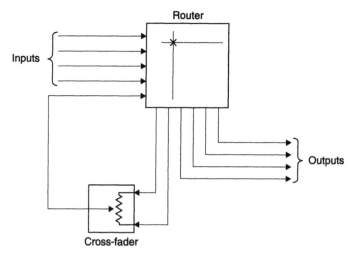

Figure 5.6 A hot cutting router can be made by assigning to it one or more cross-fader units. The input not selected by the fader can be switched, then the fader is operated.

5.4 Synchronizing

When digital audio signals are to be assembled from a variety of sources, either for mixing down or for transmission through a TDM router, the samples from each source must be synchronized to one another in both frequency and phase. Each source of samples must be fed with a reference sampling rate from some central generator, and will return samples at that rate. The same will be true if digital audio is being used in conjunction with VTRs. As the scanner speed and hence the audio block rate is locked to video, it follows that the audio sampling rate must be locked to video. Such a technique has been used since the earliest days of television in order to allow vision mixing, but now that audio is conveyed in discrete samples, these too must be genlocked to a reference for most production purposes.

AES11-1991 [5.1] documented standards for digital audio synchronization and requires professional equipment to be able to genlock either to a separate reference input or to the sampling rate of an AES/EBU input.

As the interface uses serial transmission, a shift register is required in order to return the samples to parallel format within equipment. The shift register is generally buffered with a parallel loading latch which allows some freedom in the exact time at which the latch is read with respect to the serial input timing.

Accordingly the standard defines synchronism as an identical sampling rate, but with no requirement for a precise phase relationship. Figure 5.7 shows the timing tolerances allowed. The beginning of a frame (the frame edge) is defined as the leading edge of the X preamble. A device which is genlocked must correctly decode an input whose frame edges are within ± 25% of the sample period. This is quite a generous margin, and corresponds to the timing shift due to putting about a kilometre of cable in series with a signal. In order to prevent tolerance buildup when passing through several devices in series, the output timing must be held within ± 5% of the sample period.

The reference signal may be an AES/EBU signal carrying programme material, or it may carry muted audio samples – the so-called digital audio reference signal (DARS). Alternatively it may just contain the sync patterns. The accuracy of the reference is specified in bits 0 and 1 of byte 4 of channel status (see Section 3.19). Two zeros indicates the signal is not reference grade (but some equipment may still be able to lock to it). 01 indicates a Grade 1 reference signal which is ±1 ppm accurate, whereas 10 indicates a Grade 2 reference signal which is ±10 ppm accurate. Clearly devices which are intended to lock to one of these references must have an appropriate phase-locked-loop capture range.

Increasing numbers of digital audio devices have a video input for synchronizing purposes. Video syncs (with or without picture) may be input, and a phase-locked loop will multiply the video frequency by an appropriate factor to produce a synchronous audio sampling clock.

5.5 Timing tolerance of serial interfaces

There are three parameters of interest when conveying audio down a serial interface, and these have quite different importance depending on the application. The parameters are:

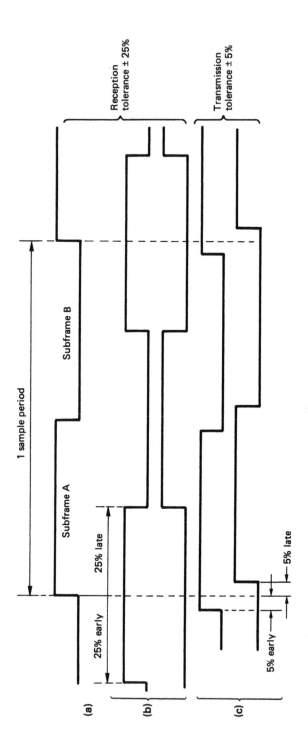

Figure 5.7 The timing accuracy required in AES/EBU signals with respect to a referene (a). Inputs over the range shown at (b) must be accepted, whereas outputs must be closer in timing to the reference as shown at (c).

a) The jitter tolerance of the serial FM data separator.
b) The jitter tolerance of the audio samples at the point of conversion back to analog.
(c) The timing accuracy of the serial signal with respect to other signals.

The serial interface is a digital interface, in that it is designed to convey discrete numerical values from one place to another. If those samples are correctly received with no numerical change, the interface is perfect. The serial interface carries clocking information, in the form of the transitions of the FM channel code and the sync patterns, and this information is designed to enable the data separator to determine the correct data values in the presence of jitter. The jitter window of the FM code is half a data bit period in the absence of noise. This becomes a quarter of a data bit when the eye opening has reached the minimum allowable in the professional specification, as can be seen from Figure 3.38. If jitter is within this limit, which corresponds to about 80 ns pk-pk the serial digital interface perfectly reproduces the sample data, irrespective of the intended use of the data. The data separator of an AES/EBU receiver requires a phase-locked loop in order to decode the serial message. This phase-locked loop will have jitter of its own, particularly if it is a digital phase-locked loop where the phase steps are of finite size. Digital phase-locked loops are easier to implement along with other logic in integrated circuits. There is no point in making the jitter of the phase-locked loop vanishingly small as the jitter tolerance of the channel code will absorb it. In fact the digital phase-locked loop is simpler to implement and locks up quicker if it has larger phase steps and therefore more jitter.

This has no effect on the ability of the interface to convey discrete values, and if the data transfer is simply an input to a digital recorder or router no other parameter is of consequence as the data values will be faithfully recorded. However, it is a further requirement in some applications that a sampling clock for a converter is derived from a serial interface signal.

The jitter tolerance of converter clocks is measured in hundreds of picoseconds. Thus a phase-locked loop in the FM data separator of a serial receiver chip is quite unable to drive a converter directly as the jitter it contains will be as much as a thousand times too great. Nevertheless this is exactly how a great many consumer outboard DACs are built, regardless of price. The consequence of this poor engineering is that the serial interface is no longer truly digital. Analog variations in the interface waveform cause variations in the converter clock jitter and thus variations in the reproduced sound quality.

Figure 5.8 shows how an outboard converter should be configured. The serial data separator has its own phase-locked loop which is less jittery than the serial waveform and so recovers the audio data. The serial data are presented to a shift register which is read in parallel to a latch when an entire sample is present by a clock edge from the data separator. The data separator has done its job of correctly returning a sample value to parallel format. A quite separate phase-locked loop with extremely high damping and low jitter is used to regenerate the sampling clock. This may use a crystal oscillator or it may be a number of loops in series to increase the order of the jitter filtering. In the professional channel status, bit 5 of byte 0 indicates whether the source is locked or unlocked. This bit can be used to change the damping factor of the phase-locked loop or to switch from a crystal to a varicap oscillator. When the source is unlocked, perhaps because a recorder is in varispeed, the capture range of the phase-locked loop can be widened and the increased jitter is accepted. When the source is locked, the capture range is reduced and the jitter is rejected.

The third timing criterion is only relevant when more than one signal is involved as it affects the ability of, for example, a mixer to combine two inputs.

In order to decide which criterion is most important, the following may be helpful. A single signal which is involved in a data transfer to a recording medium is concerned only with eye pattern jitter as this affects the data reliability.

A signal which is to be converted to analog is concerned primarily with the jitter at the converter clock. Signals which are to be mixed are concerned with the eye pattern jitter and the relative timing. If the mix is to be monitored, all three parameters become important.

5.6 Asynchronous operation

In practical situations, genlocking is not always possible. In a satellite transmission, it is not really practicable to genlock a studio complex halfway round the world to another. Outside broadcasts may be required to generate their own master timing for the same reason. When genlock is not achieved, there will be a slow slippage of sample phase between source and destination due to such factors as drift in timing generators. This phase slippage will be corrected by a synchronizer, which is intended to work with frequencies which are nominally the same. It should be contrasted with the sampling-rate converter which can work at arbitrary but generally greater frequency relationships. Although a sampling-rate converter can act as a synchronizer, it is a very expensive way of doing the job. A synchronizer can be thought of as a lower cost version of a sampling-rate converter which is constrained in the rate difference it can accept.

In one implementation of a digital audio synchronizer [5.2], memory is used as a timebase corrector. Samples are written into the memory with the frequency and phase of the source and, when the memory is half-full, samples are read out

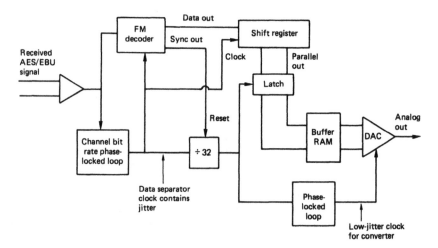

Figure 5.8 In an outboard converter, the clock from the data separator is not sufficiently free of jitter and additional clock regeneration is necessary to drive the DAC.

with the frequency and phase of the destination. Clearly if there is a net rate difference, the memory will either fill up or empty over a period of time, and in order to recentre the address relationship, it will be necessary to jump the read address. This will cause samples to be omitted or repeated, depending on the relationship of source rate to destination rate, and would be audible on programme material. The solution is to detect pauses or low level passages and permit jumping only at such times. The process is illustrated in Figure 5.9. Such synchronizers must have sufficient memory capacity to absorb timing differences between quiet passages where jumping is possible, and so the average delay introduced by them is quite large, typically 128 samples. They are, however, relatively inexpensive. An alternative to address jumping is to undertake sampling-rate conversion for a short period (Figure 5.10) in order to slip the input/output relationship by one sample [5.3]. If this is done when the signal level is low, short wordlength logic can be used.

5.7 Sound in syncs

In analog video having monophonic audio a system was developed by the BBC which allowed digital audio to be carried entirely within a video signal. This was specifically designed for distribution of programme material from studio to transmitters. The incorporation of both audio and video in one signal resulted in

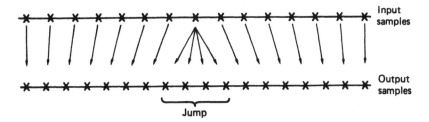

Figure 5.9 In jump synchronizing, input samples are subjected to a varying delay to align them with output timing. Eventually the sample relationship is forced to jump to prevent delay building up. As shown here, this results in several samples being repeated, and can only be undertaken during programme pauses, or at very low audio levels. If the input rate exceeds the output rate, some samples will be lost.

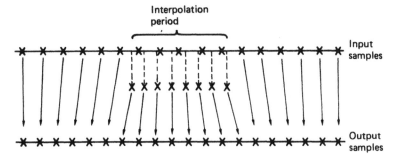

Figure 5.10 An alternative synchronizing process is to use a short period of interpolation in order to regulate the delay in the synchronizer.

a saving on land lines and, of course, it was impossible to misroute so that video could never be received without the accompanying audio.

In mono SIS, the analog audio was sampled at twice line rate and a mild compression was applied (see Section 5.14) which resulted in 10-bit samples. Twenty bits could be transmitted during sync tip by replacing the continuous sync tip voltage with a bi-level signal. A small amount of buffering was needed at each end of the SIS link in order to accept and output evenly spaced samples. The quality loss due to the mild compression factor applied was found to be insignificant in comparison to the losses and distortions suffered during analog distribution.

5.8 Analog television sound broadcasting

In order to broadcast television signals, they need to modulate a suitable radio frequency carrier. In terrestrial analog television systems, amplitude modulation (AM) is used for the video signal. Conventional AM produces symmetrical sidebands above and below the carrier frequency, effectively doubling the spectrum width required.

Conventional AM is also known as double sideband (DSB). In communications, a technique known as single sideband (SSB) exists, in which only one sideband is transmitted. This occupies half the spectrum space, but requires a complex receiver which has to regenerate the carrier frequency. This approach was considered too expensive for television receivers and so a compromise was reached in which the carrier and the upper sideband is transmitted intact, but most of the lower sideband is suppressed to leave what is called a vestigial sideband. In vestigial sideband working (VSB or AM-VSB) the receiver is no more complex than in the DSB system, but a useful spectrum saving is obtained.

Figure 5.11 shows that in addition to the vision carrier, a sound carrier is also present. This is frequency modulated to give better SNR in the audio. Thus all conventional television receivers are actually double receivers allowing simultaneous sound and picture reception. In practice the superhet television receiver has a single local oscillator which heterodynes the received vision carrier to the IF frequency of the vision receiver. This will result in an audio signal at a higher intermediate frequency than the video and so a double receiver can be obtained simply by using an additional IF strip and an FM discriminator.

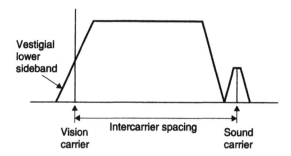

Figure 5.11 Typical transmitted television channel showing VSB video and sound carrier.

The principle of frequency modulation is shown in Figure 5.12. The nominal or centre frequency of the modulator is generated during muting, positive audio voltages increase the frequency and negative voltages reduce it. On reception the FM signal is amplified by a very large gain factor and then clipped, so that amplitude variations are largely eliminated, helping to reduce the effect of interference.

It is a characteristic of FM systems that the noise in the channel is proportional to frequency. Stated differently, the signal-to-noise ratio is triangular. Figure 5.13 shows why this should be so. An ideal demodulator removes amplitude disturbances by clipping the signal. The demodulator works by analysing the position of zero crossings in the FM signal with respect to the time axis. Noise added to a sloping signal can change the position of a zero crossing, and cause the estimated frequency to be in error, resulting in noise in the demodulated baseband signal. Figure 5.13 shows that as frequency rises, the period of a signal becomes smaller. As a result, a given disturbance on the time axis assumes a greater proportion of the signal period at high frequencies than it does at low frequencies. As a result, the SNR deteriorates at 6 dB per octave.

In FM audio broadcasting it is usual to combat the triangular noise by using pre-emphasis. Figure 5.14 shows that prior to transmission, a high frequency boost is applied with a specified time constant. After demodulation, an HF cut with the same time constant is applied, resulting in a flat overall frequency response. The reduced HF gain at the receiver reduces the level of HF noise. The use of pre-emphasis assumes that the treble energy in programme material is less than at other frequencies. If this is not so, the use of pre-emphasis will cause premature overmodulation.

In several countries the analog sound transmission system has been converted to offer stereo or dual language. In countries having sufficient spectral space around the mono sound carrier this was simply converted to an FM multiplex stereo carrier using the same principles as are employed in FM stereo radio. In an FM stereo multiplex [5.4], the existing mono signal becomes L + R and existing mono receivers will decode it as such. An additional L − R or S signal is derived and used to modulate a subcarrier of typically twice line rate prior to FM modulation. Noise reduction is often used in the S encoder and receivers need a suitable decoder.

Analog multiplex systems generally have too much crosstalk for dual language operation and some countries have adopted systems in which there are simply two FM carriers: the original mono carrier and a further carrier at a higher

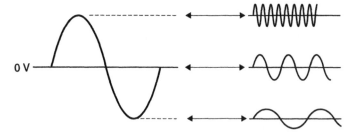

Figure 5.12 In frequency modulation, the audio signal voltage determines the frequency of the carrier. Decoding the frequency only will reduce interference which primarily affects the amplitude of the signal.

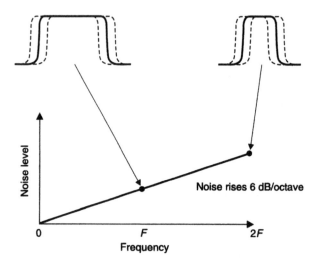

Figure 5.13 Noise disturbs the position of zero crossings. In an FM signal the effect forms a greater proportion of the cycle as frequency rises, hence the triangular noise spectrum.

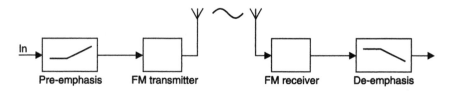

Figure 5.14 In FM, pre-emphasis is used prior to modulation to combat the triangular noise characteristic. All receivers need an opposing de-emphasis step.

frequency. The mono carrier transmits L + R, but the second carrier only sends R. In stereo receivers a simple subtraction results in L, whereas in mono receivers only L + R is obtained. The reason for sending R in the second channel is that in dual language mode the first channel becomes the primary language and the second channel becomes the secondary language, minimizing crosstalk.

In conventional terrestrial broadcasting, channels can be re-used by transmitters located a certain distance apart, but midway the channel cannot be used at all because it contains an overlapping mess; it becomes a taboo channel in that area. There is thus economic pressure on broadcasters to transmit programmes using spectrum more efficiently and to find ways of eliminating taboo channels. The only approach for the future is to use digital techniques.

5.9 Embedded audio

The serial digital interface (SDI) is becoming increasingly popular in television production installations for routing digital video signals. It exists in component and composite versions which differ primarily in the bit rate employed. In component SDI, there is provision for ancillary data packets to be sent during blanking [5.5, 5.6]. The high clock rate of component means that there is capacity for up to 16 audio channels sent in four groups. Composite SDI has to convey the digitized analog sync edges and bursts and only sync tip is available for ancillary data. As a result of this and the lower clock rate composite has much less capacity for ancillary data than component, although it is still possible to transmit one audio data packet carrying four audio channels in one group. Figure 5.15(a) shows where the ancillary data may be located for PAL and Figure 5.15(b) shows the locations for NTSC.

As was shown in Section 3.19, the data content of the AES/EBU digital audio subframe consists of Valid, User and Channel status bits, a 20-bit sample and 4 auxiliary bits which optionally may be appended to the main sample to produce a 24-bit sample. The AES recommends sampling rates of 48, 44.1 and 32 kHz, but the interface permits variable sampling rates. SDI has various levels of support for the wide range of audio possibilities and these levels are defined in Figure 5.16. The default or minimum level is Level A which operates only with a video-synchronous 48 kHz sampling rate and transmits V, U, C and the main 20-bit

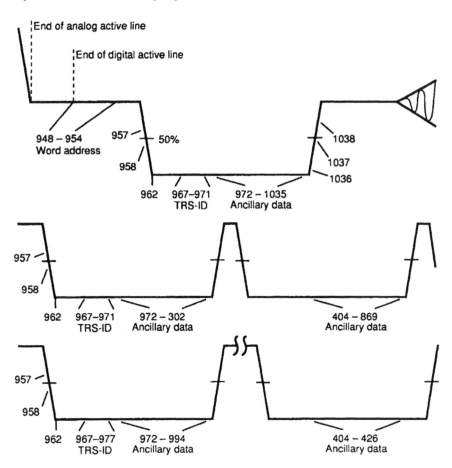

Figure 5.15(a) Ancillary data locations for PAL.

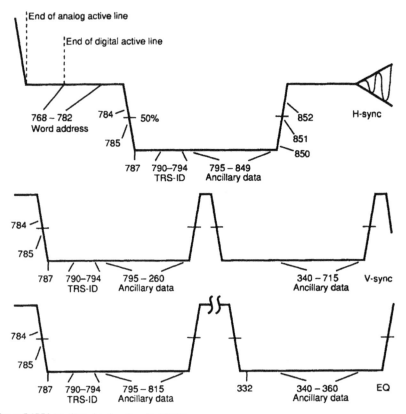

Figure 5.15(b) Ancillary data locations for NTSC.

sample only. As Level A is a default it need not be signalled to a receiver as the presence of IDs in the ancillary data is enough to ensure correct decoding. However, all other levels require an audio control packet to be transmitted to teach the receiver how to handle the embedded audio data. The audio control packet is transmitted once per field in the second horizontal ancillary space after the video switching point before any associated audio sample data. One audio control packet is required per group of audio channels.

If it is required to send 24-bit samples, the additional 4 bits of each sample are placed in extended data packets which must directly follow the associated group of audio samples in the same ancillary data space.

There are thus three kinds of packet used in embedded audio: the audio data packet which carries up to four channels of digital audio, the extended data packet and the audio control packet.

In component systems, ancillary data begin with a reversed TRS or sync pattern. Normal video receivers will not detect this pattern and so ancillary data cannot be mistaken for video samples. The ancillary data TRS consists of all zeros followed by all ones twice. There is no separate TRS for ancillary data in composite. Immediately following the usual TRS, there will be an ancillary data flag whose value must be $3FC_{16}$. Following the ancillary TRS or data flag is a data ID

	A (Default)	Synchronous 48 kHz, 20 bit audio, 48 sample buffer
	B	Synchronous 48 kHz for composite video, 64 sample buffer to receive
		20 bits from 24 bit data
	C	Synchronous 48 kHz 24 bit with extended packets
	D	Asynchronous audio
	E	44.1 kHz audio
	F	32 kHz audio
	G	32–48 kHz variable sampling rate
	H	Audio frame sequence
	I	Time delay tracking
	J	Non-coincident channel status Z bits in a pair

(left margin label, spanning rows: Needs audio control packet)

Figure 5.16 The different levels of implementation of embedded audio. Level A is default.

word which contains one of a number of standardized codes which tell the receiver how to interpret the ancillary packet. Figure 5.17 shows a list of ID codes for various types of packets. Next come the data block number and the data block count parameters. The data block number increments by 1 on each instance of a block with a given ID number. On reaching 255 it overflows and recommences counting. Next, a data count parameter specifies how many symbols of data are being sent in this block. Typical values for the data count are 36_{10} for a small packet and 48_{10} for a large packet. These parameters help an audio extractor to assemble contiguous data relating to a given set of audio channels.

Figure 5.18 shows the structure of the audio data packing. In order to prevent accidental generation of reserved synchronizing patterns, bit 9 is the inverse of bit 8 so the effective system wordlength is 9 bits. Three 9-bit symbols are used to convey all of the AES/EBU subframe data except for the 4 auxiliary bits. Since four audio channels can be conveyed, there are two 'Ch' or channel number bits which specify the audio channel number to which the subframe belongs. A further bit, Z, specifies the beginning of the 192 sample channel status message. V, U and C have the same significance as in the normal AES/EBU standard, but the

	Group 1	Group 2	Group 3	Group 4
Audio data	2FF	1FD	1FB	2F9
Audio CTL	1EF	2EE	2ED	1EC
Ext. data	1FE	2FC	2FA	1F8

Figure 5.17 The different packet types have different ID codes as shown here.

Address / Bit	x3	x3 + 1	x3 + 2
B9	$\overline{B8}$	$\overline{B8}$	$\overline{B8}$
B8	A (2^5)	A (2^{14})	P
B7	A (2^4)	A (2^{13})	C
B6	A (2^3)	A (2^{12})	U
B5	A (2^2)	A (2^{11})	V
B4	A (2^1)	A (2^{10})	A MSB (2^{19})
B3	A LSB (2^0)	A (2^9)	A (2^{18})
B2	CH (MSB)	A (2^8)	A (2^{17})
B1	CH (LSB)	A (2^7)	A (2^{16})
B0	Z	A (2^6)	A (2^{15})

Figure 5.18 AES/EBU data for one audio sample is sent as three 9-bit symbols. A = audio sample. Bit Z = AES/EBU channel status block start bit.

P bit reflects parity on the three 9-bit symbols rather than the AES/EBU definition. The three-word sets representing an audio sample will then be repeated for the remaining three channels in the packet but with different combinations of the Ch bits.

One audio sample in each of the four channels of a group requires 12 video sample periods and so packets will contain multiples of 12 samples. At the end of the packet a checksum is calculated on the entire packet contents.

If 24-bit samples are required, extended data packets must be employed in which the additional 4 bits of each audio sample in an AES/EBU frame are assembled in pairs according to Figure 5.19. Thus for every 12 symbols conveying the four 20-bit audio samples of one group in an audio data packet, two extra symbols will be required in an extended data packet.

The audio control packet structure is shown in Figure 5.20. Following the usual header are symbols representing the audio frame number, the sampling rate, the active channels, the processing delay and some reserved symbols. The sampling rate parameter allows the two AES/EBU channel pairs in a group to have different sampling rates if required. The active channel parameter simply describes which channels in a group carry meaningful audio data. The processing delay parameter denoted the delay the audio has experienced measured in audio sample periods. The parameter is a 26-bit two's complement number requiring three symbols for each channel. Since the four audio channels in a group are generally channel pairs, only two delay parameters are needed. However, if four independent channels are used, one parameter each will be required. The e bit denotes whether four individual channels or two pairs are being transmitted.

The frame number parameter comes about in 525-line systems because the frame rate is 29.97 Hz not 60 Hz. The resultant frame period does not contain a whole number of audio samples. An integer ratio is only obtained over a multiple frame sequence which is shown in Figure 5.21. The frame number conveys the position in the frame sequence. At 48 kHz odd frames hold 1602 samples and even frames hold 1601 samples in a 5-frame sequence. At 44.1 and 32 kHz the relationship is not so simple and to obtain the correct number of samples in the sequence certain frames (exceptions) have the number of samples altered. At 44.1 kHz the frame sequence is 100 frames long, whereas at 32 kHz it is 15

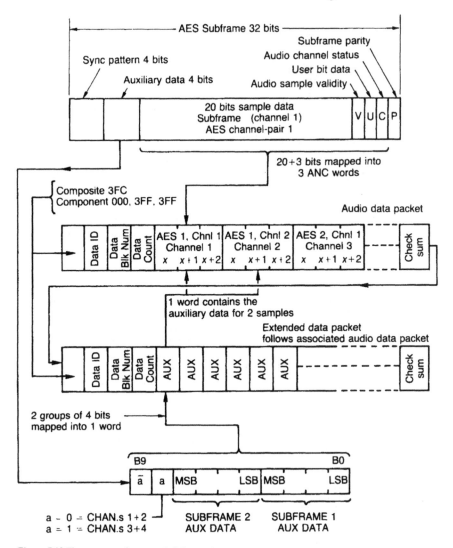

Figure 5.19 The structure of an extended data packet.

frames long. As the two-channel pairs in a group can have different sampling rates, two frame parameters are required per group. In 50 Hz systems all three sampling rates allow an integer number of samples per frame and so the frame number is irrelevant.

As the ancillary data transfer is in bursts, it is necessary to provide a little RAM buffering at both ends of the link to allow real time audio samples to be time compressed up to the video bit rate at the input and expanded back again at the receiver. Figure 5.22 shows a typical audio insertion unit in which the FIFO buffers can be seen. In such a system all that matters is that the average audio data rate is correct. Instantaneously there can be timing errors within the range of the buffers. Audio data cannot be embedded at the video switch point or in the

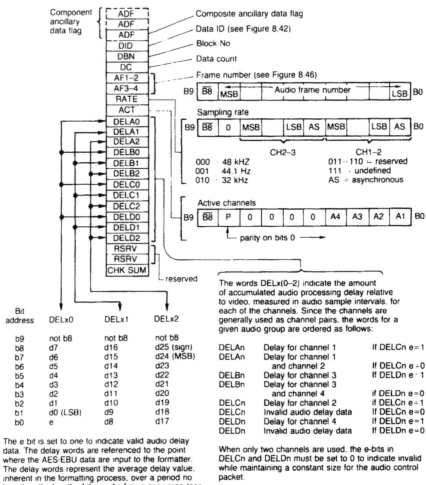

Component ancillary data flag
- ADF
- ADF
- ADF
- DID
- DBN
- DC
- AF1–2
- AF3–4
- RATE
- ACT
- DELA0
- DELA1
- DELA2
- DELB0
- DELB1
- DELB2
- DELC0
- DELC1
- DELC2
- DELD0
- DELD1
- DELD2
- RSRV
- RSRV
- CHK SUM

Composite ancillary data flag

Data ID (see Figure 8.42)

Block No

Data count

Frame number (see Figure 8.46)

B9 | B8 | MSB | — Audio frame number — | LSB | B0

Sampling rate

B9 | B8 | 0 | MSB | | LSB | AS | MSB | | LSB | AS | B0

CH2–3   CH1–2

000 · 48 kHZ	011···110 ─ reserved
001 · 44.1 Hz	111 · undefined
010 · 32 kHz	AS = asynchronous

Active channels

B9 | B8 | P | 0 | 0 | 0 | 0 | A4 | A3 | A2 | A1 | B0

└─ parity on bits 0 ──────►

reserved

Bit address	DELx0	DELx1	DELx2
b9	not b8	not b8	not b8
b8	d7	d16	d25 (sign)
b7	d6	d15	d24 (MSB)
b6	d5	d14	d23
b5	d4	d13	d22
b4	d3	d12	d21
b3	d2	d11	d20
b2	d1	d10	d19
b1	d0 (LSB)	d9	d18
b0	e	d8	d17

The e bit is set to one to indicate valid audio delay data. The delay words are referenced to the point where the AES·EBU data are input to the formatter. The delay words represent the average delay value. inherent in the formatting process, over a period no less than the length of the audio frame sequence (see Figure 8.46) plus any preexisting audio delay. Positive values indicate that the video leads the audio.

The words DELx(0–2) indicate the amount of accumulated audio processing delay relative to video. measured in audio sample intervals. for each of the channels. Since the channels are generally used as channel pairs, the words for a given audio group are ordered as follows:

DELAn	Delay for channel 1	If DELCn e=1
DELAn	Delay for channel 1 and channel 2	If DELCn e=0
DELBn	Delay for channel 3	If DELDn e=1
DELBn	Delay for channel 3 and channel 4	if DELDn e=0
DELCn	Delay for channel 2	if DELCn e=1
DELCn	Invalid audio delay data	If DELCn e=0
DELDn	Delay for channel 4	If DELDn e=1
DELDn	Invalid audio delay data	If DELDn e=0

When only two channels are used. the e-bits in DELCn and DELDn must be set to 0 to indicate invalid while maintaining a constant size for the audio control packet.

The format for the audio delay is 26 bit two's complement.

Figure 5.20 The structure of an audio control packet.

areas reserved for EDH packets, but provided that data are evenly spread throughout the frame, 20-bit audio can be embedded and retrieved with about 48 audio samples of buffering. If the additional 4 bits per sample are sent this requirement rises to 64 audio samples. The buffering stages cause the audio to be delayed with respect to the video by a few milliseconds at each insertion. Whilst this is not serious, Level I allows a delay tracking mode which allows the embedding logic to transmit the encoding delay so a subsequent receiver can compute the overall delay. If the range of the buffering is exceeded for any reason, such as a non-synchronous audio sampling rate fed to a Level A encoder, audio samples are periodically skipped or repeated in order to bring the delay under control.

It is permitted for receivers which can only handle 20-bit audio to discard the 4-bit sample extension data. However, the presence of the extension data requires

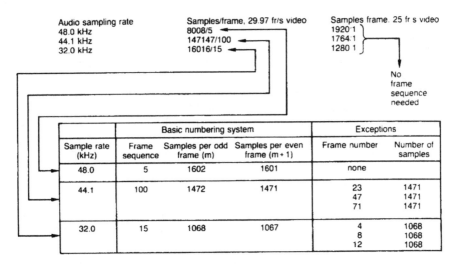

Figure 5.21 The origin of the frame sequences in 525-line systems.

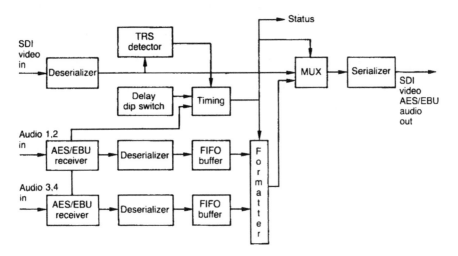

Figure 5.22 A typical audio insertion unit. See text for details.

more buffering in the receiver. A device having a buffer of only 48 samples for Level A working could experience an overflow due to the presence of the extension data.

In 48 kHz working, the average number of audio samples per channel is just over three per video line. In order to maintain the correct average audio sampling rate, the number of samples sent per line is variable and not specified in the standard. In practice a transmitter generally switches between packets containing three samples and packets containing four samples per channel per line as required, to keep the buffers from overflowing. At lower sampling rates either smaller packets can be sent or packets can be omitted from certain lines.

As a result of the switching, ancillary data packets in component video occur mostly in two sizes. The larger packet is 55 words in length, of which 48 words are data. The smaller packet contains 43 words, of which 36 are data. There is space for two large packets or three small packets in the horizontal blanking between EAV and SAV.

A typical embedded audio extractor is shown in Figure 5.23. The extractor recognizes the ancillary data TRS or flag and then decodes the ID to determine the content of the packet. The group and channel addresses are then used to direct extracted symbols to the appropriate audio channel. A FIFO memory is used to timebase expand the symbols to the correct audio sampling rate.

5.10 Introduction to compression

Compression techniques have been around for a long time and have been used in many different industries. However, compression requires complex processing and its widespread adoption had to wait until the availability of low cost digital LSI chips. Once these had been developed, compression products appeared virtually overnight. However, compression does not have to be digital; it has been seen in the analog domain as well.

Compression is a technique which allows an audio signal to be sent down a channel of restricted information capacity with less impairment than it would suffer without compression [5.7]. The use of compression can either:

(a) increase the dynamic range of the overall system beyond that allowed by the channel alone, or
(b) keep the dynamic range the same but allowing the use of a more economical channel.

The Dolby systems described in Section 4.7 are an example of the first approach.

Most digital compression systems come into the second category. Audio is converted to conventional PCM having a bit rate equal to the product of the

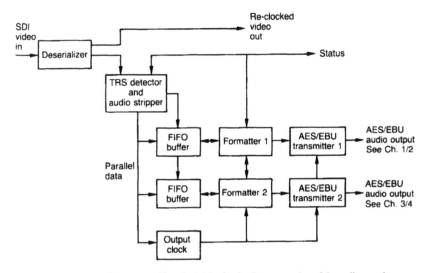

Figure 5.23 A typical audio extractor. Note the FIFOs for timebase expansion of the audio samples.

sampling rate and the wordlength. Compression is used to reduce this bit rate. The reduced bit rate offers a number of advantages:

(a) a channel of reduced bandwidth can be used for transmission, e.g. digital broadcasting
(b) the playing time of a given storage medium is extended in proportion to the compression factor, e.g. disks
(c) the cost of storing a given length of programme falls in proportion to the compression factor, e.g. archiving
(d) the access time and transfer rate of a storage medium falls in proportion to the compression factor, e.g. tape recording.

Compression works by identifying the meaningful information in the audio. This is known as the entropy. Entropy is that part of a signal which was surprising or unpredictable. The remainder is predictable and is called redundancy. Take the case of a sinewave. All cycles look the same so it is totally predictable and therefore carries no information. This follows from the fact that it has zero bandwidth.

Information in audio is unpredictable: transients for example. Figure 5.24 shows a simple example. The information capacity of a normal PCM system can be represented by an area which is the product of the bandwidth and the wordlength. The full area is transmitted at all times in a PCM system, whereas only part of it is occupied by entropy. The remainder is redundancy. If a hypothetical ideal compressor is used which can perfectly separate the entropy from the redundancy, only the entropy need be sent and the quality will not be impaired. However, if less than the entropy is sent, impairment is inevitable. It follows from this that for all audio sources there will be a critical compression factor below which there is no perceivable loss of quality. Going beyond this factor is bound to cause quality loss. In MPEG-2 a bit rate of 384 kilobits/s for stereo is generally considered to be transparent. This corresponds to a compression factor of 4:1 with respect to 48 kHz 16-bit PCM.

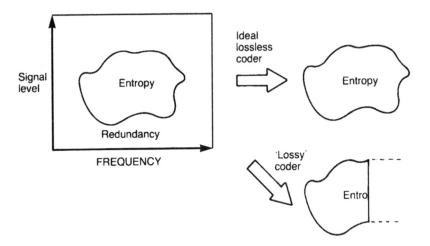

Figure 5.24 A perfect coder removed only the redundancy from the input signal and results in subjectively lossless coding. If the remaining entropy is beyond the capacity of the channel, some of it must be lost and the codec will then be lossy. An imperfect coder will also be lossy as it fails to keep all entropy.

In practice an ideal compressor will be infinitely complex and real compressors will not be able to identify the redundancy perfectly. Consequently the compression factor has to be reduced to give a safety margin. The simpler (and cheaper) the compression equipment, the larger the safety margin has to be. This is why ISO/MPEG audio compression is available in three levels: each gives a different balance between cost/complexity and compression factor.

Compression would not be possible were it not for the phenomenon of auditory masking which is the reduced ability to detect one sound in the presence of another. The threshold of human hearing is a function of frequency and reaches a minimum in the speech band as shown in Figure 5.25(a). In the presence of a single tone, shown in Figure 5.25(b), the threshold is raised over a range of frequencies, most noticeably above the frequency of the masker. This is due to the finite Q-factor of the basilar membrane. Any signal below the new threshold will not be heard. This is why tape hiss is never heard during a loud passage in a recording. A compressor has to analyse the input audio spectrum and determine the masked threshold of hearing from time to time. Once this is done the noise floor of the audio can be raised, for example, by reducing the sample wordlength. Provided the increased noise does not rise above the masked threshold it will be inaudible.

It is necessary to use caution when compression is to be employed with stereo audio. In a stereo image, sound sources may be spatially separated from one another. When this is the case masking is much less effective, as evidenced by the 'cocktail party effect' in which it is possible to concentrate on one conversation in the presence of others. Consequently when listening in stereo with high quality loudspeakers, defects can be detected in the most advanced compression algorithms which disappear when the signals are summed to mono. Correct stereo compression will require the compressor correctly to analyse the spatial location of sound sources prior to determining the masking model. The compression factors achieved will be lower simply because the spatial information in stereo is entropy which has to be transmitted.

Four major tools are used in compression. Each alone can only deliver a certain amount of compression, but used in combination higher compression factors can be achieved.

Companding reduces the dynamic range of the input signal just like a Dolby system. It can then be sent through a noisier system, i.e. one with shorter wordlength. In two's complement audio coding, every 6 dB reduction in level from peak means that a high order bit copies the sign bit. This is the phenomenon of sign extension. Consequently at low levels, high order bits in the samples are completely predictable and need not be sent. For example at -24 dBFs there are four redundant bits in each sample and a lossless compression is possible by removing them. However, at high levels, bits must be removed from the low order end of the samples. This effectively raises the size of the quantizing step, hence the term requantizing. The consequence is quantizing error which is only allowable if it is masked by higher programme level.

Sub-band coding splits the audio spectrum into a number of smaller bands. In much programme material most energy is present in only a few bands and the remainder contain relatively low levels, thus companding is particularly useful when combined with bandsplitting. The noise floor can be individually raised in each band in order to approximate more closely the masking threshold as Figure

5.25(c) shows. Figure 5.25(c) also introduces the concept of the noise-to-masking ratio (NMR) which is a measure of the tolerance band between the noise floor and the masking threshold. In all quality systems a positive NMR must be maintained.

Transform coding converts time domain signals into frequency domain coefficients. This in itself does not achieve any compression, but in the frequency domain, the spectrum often changes quite slowly and the coefficients do not need to be sent very often. Clearly this approach falls down in the presence of transients and an adaptive approach will be needed.

Predictive coding uses a matched pair of predicting circuits at encoder and decoder as shown in Figure 5.26. A predictor is basically a device which looks at a run of previous samples and from them attempts to establish the value of the next sample. Think of it as a lop-sided interpolator. The prediction is then subtracted from the actual input sample to determine the prediction error which is actually transmitted. The decoder makes the same prediction and adds the error to correct it. Predictive coders have the advantage that they use the history of the signal and can therefore operate with only a small processing delay. However they cannot, of course, see transients coming and these cause very large errors. In practice the error needs to be companded to handle transients.

Compressors have to identify and transmit the unpredictable part of an input signal. Unfortunately noise is also unpredictable and a compressor cannot tell noise from entropy and will attempt to code it. This uses up valuable bit rate which is not then available for the genuine audio. Paradoxically, then, signals

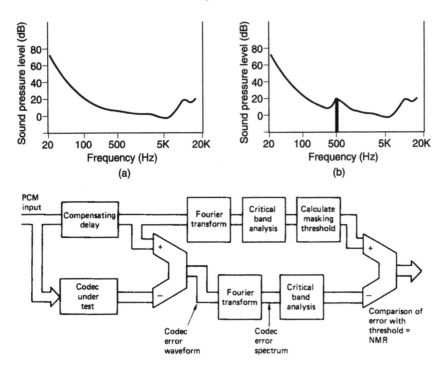

Figure 5.25 (a) The threshold of hearing is modified by a tone (b). The area under the elevated threshold is masked. (c) The noise-to-masking ratio is derived as shown here.

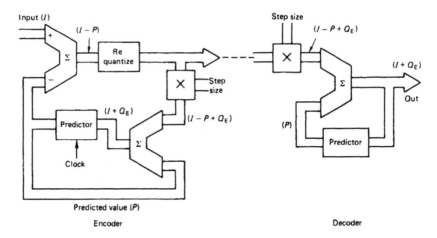

Figure 5.26 A predictive coder resembles the operation of a sigma-delta modulator which has a digital input. See text for details.

which are destined to be compressed must be created to a very high standard with a low noise floor.

Compressors actually raise the noise floor. Consequently a second compressor will have difficulty due to the noise caused by the first compressor. Thus cascading codecs is a bad idea and if it must be done a comfortable NMR must be maintained on the first compression.

Many of today's quality audio compression systems use the principle of sub-band coding in which the audio spectrum is divided up into a large number of frequency bands. One method which is often used is the quadrature mirror (QMF) which converts a PCM sample stream into two sample streams of half the input sampling rate, so that the output data rate is equal to the input data rate. The frequencies in the lower half of the audio spectrum are carried in one sample stream, and those in the upper half of the spectrum are carried in the other. The lower frequency output is a PCM band-limited representation of the input waveform and can be reconstructed with a low pass filter. However, the upper frequency output is down-sampled or aliased and can only be reconstructed by an interpolator having a bandpass response, also known as a synthesis filter. As only signals within the passband can be output, the original waveform will result as the aliased version lies outside the passband. The audio band can be split into as many bands as required by cascading QMFs in a tree.

Figure 5.27 shows that in ISO-MPEG audio compression 32 sub-bands are used. The decomposed sub-band data are then assembled into blocks of 12 samples each, prior to compression. Sub-band blocks are also referred to as 'frequency bins'.

Coding in each bin is by a combination of gain ranging and requantizing. The gain ranging is achieved by multiplying the sample values by a scale factor and requantizing is by rounding off LSBs to the required wordlength. For example, if the waveform amplitude is 36 dB less than maximum, there will be at least 6 bits in each sample which copy the sign bit. Multiplying by 64, the equivalent of a 6-bit shift, will bring the high order bits of the sample into use, allowing LSBs to

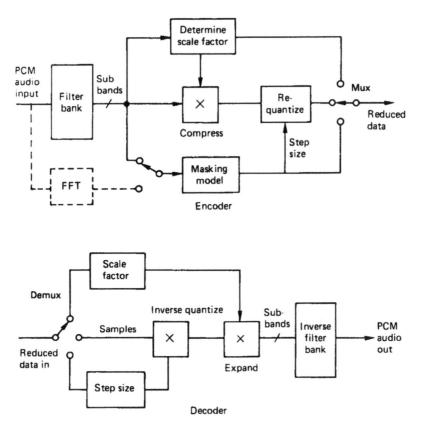

Figure 5.27 A simple sub-band coder. The bit allocation may come from analysis of the sub-band energy, or, for greater reduction, from a spectral analysis in a side chain.

be lost by rounding. The shorter the wordlength, the greater the coding gain, but the coarser the quantization steps and therefore the level of quantization error. In this example, rounding off 6 bits would not affect the signal quality at all. Losing more than 6 bits could raise the noise level compared to that in the input signal and can only be done if that noise will be masked. In addition to the companded requantized samples, the scale factor must be sent so that the decoder can divide by the correct value to get the level correct.

A fixed compression factor will set the overall size of the coded output block. Some bins can have long wordlengths if others have short wordlengths. The process of sharing wordlengths between the sub-bands is known as bit allocation. To help the decoder to deserialize the bitstream, it will be necessary to transmit an allocation table specifying the wordlength in each bin. The allocation process compares the spectral content of the input with an auditory masking model to determine the degree of masking which is taking place. Where masking takes place, more LSBs can be removed provided the quantizing noise remains below the masking level. The bit allocation may be performed crudely by analysing the power in each sub-band, or by a side chain which performs a spectral analysis or transform of the audio. The bit allocation may be iterative as wordlengths are traded to obtain the best result. The compression factor is controlled by changing the block size parameter. The bit allocator simply iterates until the new block size is filled. The decoder will automatically deserialize any size block using the allocation table. Consequently it is easy to change the compression factor for different jobs without modifying the hardware.

The ISO-MPEG Layer I system is designed to be simple and inexpensive and so the sub-bands themselves are used as a spectral analysis of the input in order to determine the bit allocation. In Layer II the same filterbank as Layer I is used, and the blocks are the same size. The same block companding scheme is used. A side chain FFT performs an analysis of the audio spectrum eight times more accurately than the sub-band filter. This allows higher compression factors to be obtained. With a higher compression factor, the allocation table and the scale factors form an increasing proportion of the message. The scale factor of successive blocks in the same band differs by 2 dB less than 10% of the time, and advantage is taken of this redundancy by analysing sets of three successive scale factors. On stationary programme, only one scale factor out of three is sent. As transient content increases in a given sub-band, two or three scale factors will be sent.

Layer III is the most complex layer of the ISO-MPEG standard, and is only really necessary when very high compression factors must be used. It is not considered here as it is unlikely to be used for television sound.

A compressed transmission consists of samples in bins and a scale factor and wordlength relating to each bin. Some synchronizing means is needed to allow the beginning of the block to be identified. Demultiplexing is done using the allocation data. Once all of the samples are back in their respective frequency bins, the level of each bin is returned to the original value. This is achieved by using the scale factor to oppose the gain applied in the coder. The sub-bands can then be recombined into a continuous audio spectrum in the output filter which produces conventional PCM.

Frequency analysis or transform coding allows any repetitive waveform to be represented by coefficients describing a set of harmonically related components of suitable amplitude and phase. The transform of a typical audio waveform changes relatively slowly. The slow speech of an organ pipe or a violin string, or the slow decay of most musical sounds, allow the rate at which the transform is sampled to be reduced, and a coding gain results. Not all frequencies are simultaneously in real audio, and so not all coefficients need be sent. A further coding gain will be achieved if the components which will experience masking are quantized more coarsely.

Transforms require blocks of samples and a practical solution, used for example in Dolby AC-3, is to cut the waveform into short overlapping windows and then to transform each individually as shown in Figure 5.28. Every input sample appears in two windows with variable weighting depending upon its position along the time axis.

The DFT (discrete frequency transform) requires intensive computation, owing to the requirement to use complex arithmetic to render the phase of the components as well as the amplitude. An alternative is to use the discrete cosine transform (DCT). In the DCT a time-mirrored version of the window block is placed in front of the original block before the transform process. All sinusoidal components then cancel out and only cosine coefficients are computed. In the modified discrete cosine transform (MDCT), windows with 50% overlap are used, producing twice as many coefficients as necessary. These are sub-sampled

Figure 5.28 Transform coding can only be practically performed on short blocks. These are overlapped using window functions in order to handle continuous waveform.

by a factor of two to give a critically sampled transform, which results in potential aliasing in the frequency domain. However, by making a slight modification to the transform, the alias products in the second half of a given window are equal in size but of opposite polarity to the alias products in the first half of the next window, and so will be cancelled on reconstruction. This is the principle of time domain aliasing cancellation (TDAC).

The coefficient requantizing in the coder raises the quantizing noise in the frequency bin, but it does so over the entire duration of the block. Figure 5.29 shows that if a transient occurs towards the end of a block, the decoder will reproduce the waveform correctly, but the quantizing noise will start at the beginning of the block and may result in a pre-echo where the noise is audible before the transient.

This can be avoided using a variable time window according to the transient content of the audio waveform. When musical transients occur, short blocks are necessary and the frequency resolution and hence the coding gain will be low. At other times the blocks become longer and the frequency resolution of the transform rises, allowing a greater coding gain.

The transform of an audio signal is computed in the main signal path in a transform coder, and has sufficient frequency resolution to drive the masking model directly.

5.11 Introduction to NICAM 728

This system was developed by the BBC to allow the two additional high quality digital sound channels to be carried on terrestrial television broadcasts. The

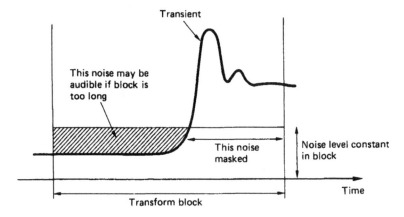

Figure 5.29 If a transient occurs towards the end of a transform block, the quantizing noise will still be present at the beginning of the block and may result in a pre-echo where the noise is audible before the transient.

introduction of stereo sound with television could not be at the expense of incompatibility with the existing monophonic sound channel. In NICAM 728 an additional low power subcarrier is positioned just above the analog sound carrier, which is retained. The relationship is shown in Figure 5.30. The power of the digital subcarrier is about one-hundredth that of the main vision carrier, and so existing monophonic receivers will reject it.

Since the digital carrier is effectively shoehorned into the gap between television channels, it is necessary to ensure that the spectral width of the intruder is minimized to prevent interference. As a further measure, the power of the existing audio carrier is halved when the digital carrier is present.

Figure 5.31 shows the stages through which the audio must pass. The audio sampling rate used is 32 kHz which offers similar bandwidth to that of an FM stereo radio broadcast. Samples are originally quantized to 14-bit resolution in two's complement code. From an analog source this causes no problem, but from a professional digital source having longer wordlength and higher sampling rate it would be necessary to pass through a rate converter, a digital equalizer to provide pre-emphasis, an optional digital compressor in the case of wide dynamic range signals and then through a truncation circuit incorporating digital dither.

The 14-bit samples are block companded to reduce data rate. During each 1 ms block, 32 samples are input from each audio channel. The magnitude of the largest sample in each channel is independently assessed, and used to determine the gain range or scale factor to be used. Every sample in each channel in a given block will then be scaled by the same amount and truncated to 10 bits. An eleventh bit present on each sample combines the scale factor of the channel with parity bits for error detection. The encoding process is described as a near instantaneously companded audio multiplex, NICAM for short. The resultant data now

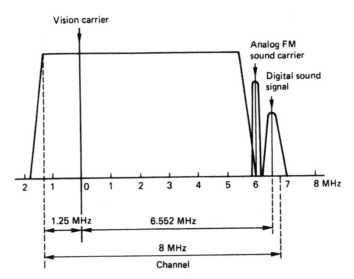

Figure 5.30 The additional carrier needed for digital stereo sound is squeezed in between television channels as shown here. The digital carrier is of much lower power than the analog signals, and is randomized prior to transmission so that it has a broad, low level spectrum which is less visible on the picture.

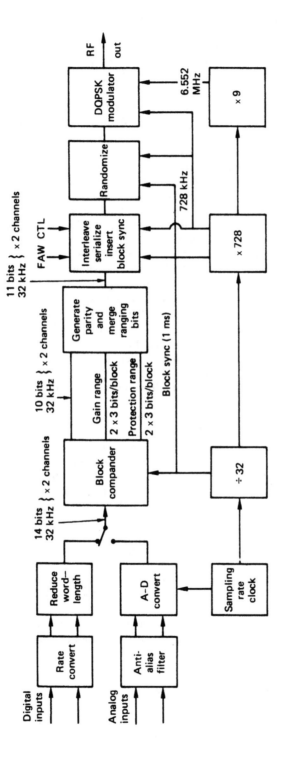

Figure 5.31 The stage necessary to generate the digital subcarrier in NICAM 728. Audio samples are block companded to reduce the bandwidth needed.

consists of $2 \times 32 \times 11 = 704$ bits per block. Bit interleaving is employed to reduce the effect of burst errors.

At the beginning of each block a synchronizing byte, known as a frame alignment word (FAW), is followed by 5 control bits and 11 additional data bits, making a total of 728 bits per frame, hence the number in the system name.

As there are 1000 frames per second, the bit rate is 728 kbits/s. In the UK this is multiplied by 9 to obtain the digital carrier frequency of 6.552 MHz, although some other countries use a different subcarrier spacing.

The carrier can be radiated in four discrete phases. If there is an absolute relation between the data and the phase, the technique is called quadrature phase shift keying or QPSK. Each period of the transmitted waveform can have one of four phases and therefore conveys the value of two data bits. In order to resist reflections in broadcasting, QPSK can be modified so that a knowledge of absolute phase is not needed at the receiver. Instead of encoding the signal phase, the data determine the magnitude of a phase shift. This is known as differential quadrature phase shift keying or DQPSK and is the modulation scheme used for NICAM 728. A DQPSK coder is shown in Figure 5.32 and as before 2 bits are conveyed for each transmitted period. It will be seen that one bit pattern results in no phase change. If this pattern is sustained the entire transmitter power will be concentrated in the carrier. This can cause patterning on the associated television pictures. A randomizing technique is used to overcome the problem. The effect is to spread the signal energy uniformly throughout the allowable channel bandwidth so that it has less energy at a given frequency. This reduces patterning on the analog video signal in addition to making the signal more resistant to multipath reception which tends to remove notches from the spectrum.

A pseudo-random sequence (prs) generator is used to generate the randomizing sequence used in NICAM. A 9-bit device has a sequence length of 511, and is preset to a standard value of all ones at the beginning of each frame. The serialized data are XORed with the LSB of the Galois field, which randomizes the output which then goes to the modulator. The spectrum of the transmission is now determined by the spectrum of the prs. This has a spikey $\sin x/x$ envelope. The frequencies beyond the first nulls are filtered out at the transmitter, leaving the characteristic 'dead hedgehog' shape seen on a spectrum analyser.

On reception, the de-randomizer must contain the identical ring counter which must also be set to the starting condition to bit accuracy. Its output is then added to the data stream from the demodulator. The randomizing will effectively then have been added twice to the data in modulo 2, and as a result is cancelled out leaving the original serial data.

Randomizing is used for everything except the FAW. On reception, the FAW is detected and used to synchronize the pseudo-random generator to restore the original data.

5.12 NICAM 728 frame structure

Figure 5.33 shows the general structure of a frame. Following the sync pattern or FAW is the application control field. The application control bits determine the

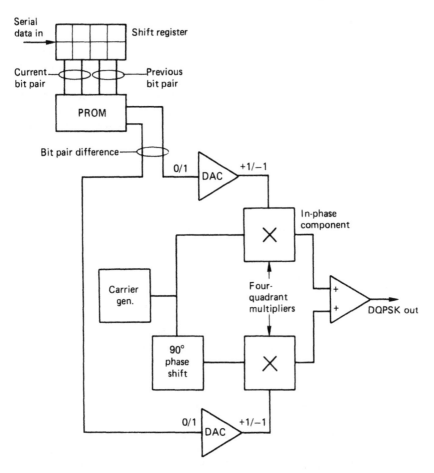

Figure 5.32 A DQPSK coder conveys 2 bits for each modulation period. See text for details.

significance of following data, which can be stereo audio, two independent mono signals, mono audio and data or data only.

Control bits C_1, C_2 and C_3 have eight combinations, of which only four are currently standardized. Current receivers are designed to mute audio if C_3 becomes 1.

The frame flag bit C_0 spends eight frames high then eight frames low in an endless 16 frame sequence which is used to synchronize changes in channel usage. In the last 16-frame sequence of the old application, the application control bits change to herald the new application, whereas the actual data change to the new application on the next 16-frame sequence.

The reserve sound switching flag, C_4, is set to 1 if the analog sound being broadcast is derived from the digital stereo. This fact can be stored by the receiver and used to initiate automatic switching to analog sound in the case of loss of the digital channels.

The additional data bits AD_0 through AD_{10} are as yet undefined, and reserved for future applications.

5.13 The NICAM sound/data block

The remaining 704 bits in each frame may be either audio samples or data. The two channels of stereo audio are multiplexed into each frame, but multiplexing does not occur in any other case. If two mono audio channels are sent, they occupy alternate frames. Figure 5.33(a) shows a stereo frame, where the A channel is carried in odd numbered samples, whereas Figure 5.33(b) shows a mono frame, where the M1 channel is carried in odd numbered frames. The format for data has yet to be defined.

The sound/data block of NICAM 728 is in fact identical in structure to the first-level protected companded sound signal block of the MAC/packet systems [5.8].

5.14 Companding, scaling and parity

Figure 5.34 shows how the companding works. The 14-bit samples are examined to extract the sample magnitude. Clearly the sign bit plays no part in determining magnitude, and so the diagram has a certain vertical symmetry. The most significant 9 of the remaining 13 bits will used, along with the sign bit, to produce a 10-bit mantissa. As there are only five shifted positions in which 9 bits can reside within 13 bits, clearly there are only five scaling factors. Three bits are required to describe five scaling factors, and 3 bits actually allows eight combinations.

In NICAM 728 there is no error correction. The channel is sufficiently reliable that errors are rare. If an error occurs, it is not corrected, but it is detected so that concealment can be used. The subjective effect of a bit error is roughly proportional to its significance in the companded sample, and so it is only necessary to detect errors in the most significant bits, as errors in the low order bits only will experience the same masking effects which allow the use of companding in the first case. In NICAM 728, the 6 most significant bits of each companded sample are protected by an additional parity bit. This works perfectly until the lowest magnitude samples occur in the fifth coding range. When the sample is 6 dB below the fifth coding range, the MSB of the companded sample will no longer be active, and when the sample is 12 dB below the fifth coding range the 2 most significant bits will not be active. It makes more sense to slide the 6 bits which are protected by parity down by 1 or 2 bits respectively, giving an improvement in SNR on quiet signals in the presence of errors of up to 12 dB. This explains the fact that although there are only five different scale factors, there are seven different protection ranges.

In order to convey the scale factor to the receiver during a stereo broadcast, 3 bits must be transmitted for each channel of each block. These data cannot be sent unprotected, as their loss would effectively prevent the scaling of an entire block. The scale factor bits are actually merged with the parity bits. In the case of the most significant scale factor bit R_2, reference to Figure 5.34(b) will show that when R_2 is 1, the parity bits of samples 1,7,13,19 etc., are inverted, but when R_2 is 0 they are left unchanged. Similarly if R_1 is 1, the parity bits of samples 3,9,15,21 etc., are inverted, and so on for bit R_0. As stated, errors in transmission are relatively few, and so on reception the presence of an apparent parity failure in bits 1,7,13,19 etc. is interpreted as R_2 being set to 1. An actual error would

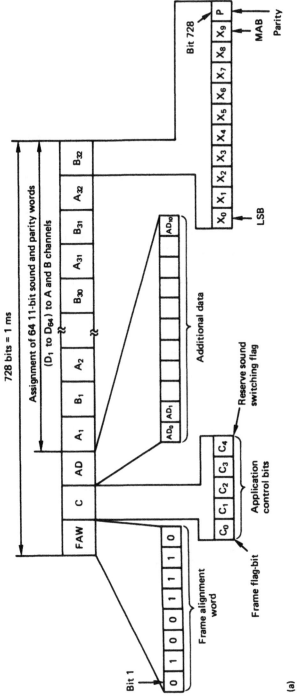

(a)

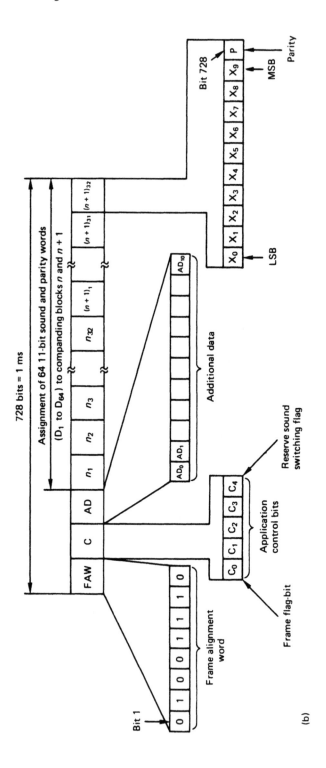

Figure 5.33 In (a), the block structure of a stereo signal multiplexes samples from both channels (A and B) into one block. In mono, shown in (b), samples from one channel only occupy a given block. The diagrams here show the data before interleaving. Adjacent bits shown here actually appear at least 16 bits apart in the data stream.

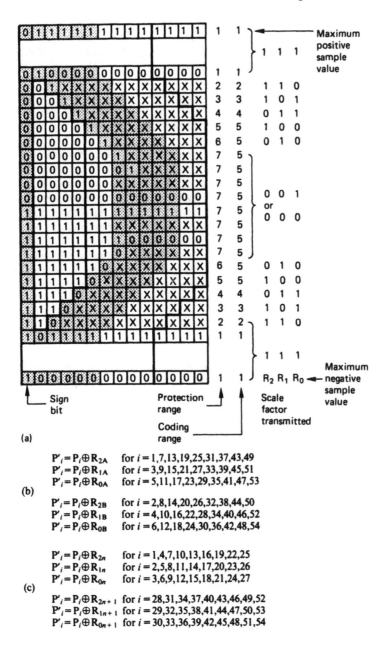

(a)

(b)

$P'_i = P_i \oplus R_{2A}$ for $i = 1,7,13,19,25,31,37,43,49$
$P'_i = P_i \oplus R_{1A}$ for $i = 3,9,15,21,27,33,39,45,51$
$P'_i = P_i \oplus R_{0A}$ for $i = 5,11,17,23,29,35,41,47,53$

$P'_i = P_i \oplus R_{2B}$ for $i = 2,8,14,20,26,32,38,44,50$
$P'_i = P_i \oplus R_{1B}$ for $i = 4,10,16,22,28,34,40,46,52$
$P'_i = P_i \oplus R_{0B}$ for $i = 6,12,18,24,30,36,42,48,54$

(c)

$P'_i = P_i \oplus R_{2n}$ for $i = 1,4,7,10,13,16,19,22,25$
$P'_i = P_i \oplus R_{1n}$ for $i = 2,5,8,11,14,17,20,23,26$
$P'_i = P_i \oplus R_{0n}$ for $i = 3,6,9,12,15,18,21,24,27$

$P'_i = P_i \oplus R_{2n+1}$ for $i = 28,31,34,37,40,43,46,49,52$
$P'_i = P_i \oplus R_{1n+1}$ for $i = 29,32,35,38,41,44,47,50,53$
$P'_i = P_i \oplus R_{0n+1}$ for $i = 30,33,36,39,42,45,48,51,54$

Figure 5.34 The companding of NICAM 728 is shown here. As the programme material becomes quieter, the high-order bits no longer play any part, since they become the same on the sign bit. In a companded sample, the sign bit and 9 other bits are transmitted, as the heavy lines indicate. There are only five gain ranges (1–5) possible. However, parity is only generated on the 6 most significant bits of the sample, except at low levels where the bits checked are slid down one or two places. Thus there are five gain ranges but seven protection ranges which are conveyed for each block by the scale factor bits R_0-R_2. Scale factor is transmitted by reversing parity at predetermined bit positions. These are shown in (b) for stereo and (c) for mono.

cause the parity check to succeed, but this would still be interpreted as an error because if, for example sample 7 had no parity error yet samples 1,13,19, 25 etc. did, it more likely that R_2 is 1 and sample 7 has a parity error than that sample 7 is correct, R_2 is 0 and all the other samples are wrong. The receiver requires majority decoding logic correctly to interpret the received parity and ranging bits. An odd number of parity checks is employed to prevent the decoder needing to toss a coin. It is possible to convey two more bits in the same way using the parity modification of samples 55 – 59 and 60 – 64, although this is not needed in the NICAM 728 application.

Figure 5.34(c) shows how the scale factor is merged with the parity on a mono transmission. In mono, each frame contains two blocks from one audio channel, and needs a separate scale factor for each. Note that some of the samples (28 – 32) from the first block are used to carry the scale factor of the second block. This causes no concern, as a frame can never be transmitted without both blocks present, and it allows samples 55 – 64 to be used for other purposes as before.

The interleaving is performed as follows: columns of an array are assembled, each of which contains four 11-bit samples. Sixteen columns are necessary to hold the 64 samples of each frame. The data bits are read out in rows and transmitted. As a result, bits in the same sample are never closer together than 16 transmitted bits.

5.15 Dual SIS

A development of the mono SIS system is in use in UK to convey NICAM digital stereo to transmitters. The problem facing the designers of DSIS was how to fit effectively twice as much data into a video sync pulse. Where an existing wide band channel having a response to DC, or at least clamping, and a good SNR is being used for digital signalling, an increase in data rate can be had using multilevel signalling or m-ary coding instead of binary. This is the basis of the DSIS coding system.

Figure 5.35 shows the four-level waveform of the UK DSIS system. Clearly the data separator must have a 2-bit ADC which can resolve the four signal levels. The gain and offset of the signal must be precisely set so that the quantizing levels register precisely with the centres of the eyes.

The DSIS system carries data which have been compressed according to the NICAM companding scheme. As no other audio information is sent to the transmitter, the latter must derive a mono signal for the existing analog mono FM channel from the DSIS data. The transmitter has a local NICAM compression decoder and DACs so that the two analog channels are available. In the case of a stereo DSIS signal, addition of L and R with a suitable gain compensation is used to derive mono. In the case of a dual language transmission this will be indicated in the DSIS status flags and the transmitter can select only the primary language for the mono carrier.

5.16 Digital broadcasting

Digital broadcasting is a technique in which the received radiation from the transmitter is decoded into discrete numerical values rather than as a continuous variable. Discrete decoding allows moderate noise levels to be rejected. As noise

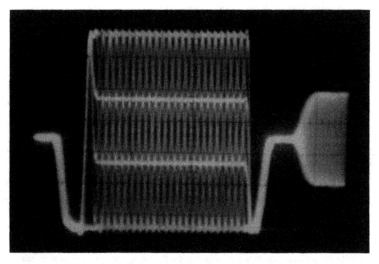

Figure 5.35 DSIS information within the TV line sync pulse.

has a distribution of levels, the higher, yet infrequent, noise levels will cause data errors, but a suitable error correction system can correct the data to the original values. In an analog system the higher amplitude noise pulses are visible in the picture and audible in the sound. The only solution is to raise the power of the transmitter. In digital systems the infrequent noise peaks are eliminated by error correction. Consequently lower transmitter power can be used in digital transmission.

In analog broadcasting, multipath transmission is a serious problem. When the direct and reflected signals are received with equal strength, it will be seen from Figure 5.36 that nulling occurs at any frequency where the path difference results in a 180-degree phase shift. Effectively a comb filter is placed in series with the signal. Clearly raising the transmitter power is of no help. A directional antenna is useful, but only in fixed installations. If the reflection is from a moving object such as a tanker ship or an aircraft, the path lengths change, and the comb response slides up and down the band. When a null passes through the station tuned in, a burst of noise is created.

Figure 5.37 shows how multiple carriers can be allocated to different programme channels on an interleaved basis. Using this technique, it will be evident that if a notch in the received spectrum occurs due to multipath cancellation this will damage a small proportion of all programmes rather than a large part of one programme. This is the spectral equivalent of physical interleaving on a disk or tape. The result is the same in that error bursts are broken up according to the interleave structure into more manageable sizes which can be corrected with less redundancy.

The ability to work in the presence of multipath cancellation is one of the great strengths of digital broadcasting. It means that several transmitters can radiate exactly the same signal so that the same channel is effectively re-used without taboo channels being necessary. With traditional modulation techniques this would be quite impossible because in certain locations between transmitters there

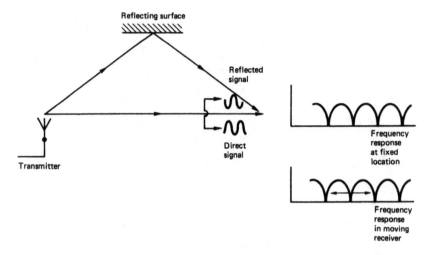

Figure 5.36 Multipath reception. When the direct and reflected signals are received with equal strength, nulling occurs at any frequency where the path difference results in a 180° phase shift.

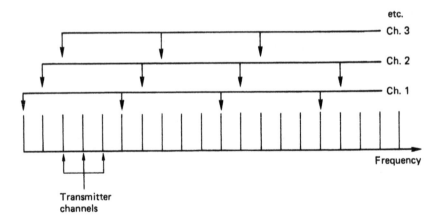

Figure 5.37 Channel interleaving is used in DAB to reduce the effect of multipath notches on a given programme.

would simply be no signal. Repeaters can be installed in shadow areas such as valleys and these can also use the same frequencies.

Where the maximum data rate in a given bandwidth is needed for economic reasons as in digital broadcasting, multi-level signalling can be combined with PSK to obtain multi-level quadrature amplitude modulation (QUAM). Figure 5.38 shows the example of 64-QUAM. Incoming 6 bit data words are split into two 3-bit words and each is used to amplitude modulate a pair of sinusoidal carriers which are generated in quadrature. The modulators are four-quadrant devices such that 2^3 amplitudes are available, four which are in phase with the carrier and four which are antiphase.

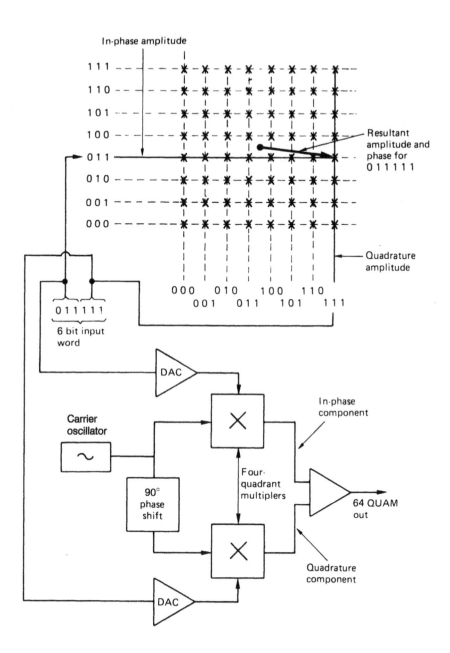

Figure 5.38 In 64-QUAM, two carriers are generated with a quadrature relationship. These are independently amplitude modulated to eight discrete levels in four quadrant multipliers. Adding the signals produces a QUAM signal having 64 unique combinations of amplitude and phase. Decoding requires the waveform to be sampled in quadrature like a colour television subcarrier.

The two AM carriers are linearly added and the result is a signal which has 2^6 or 64 combinations of amplitude and phase. The diagram showing the phase/amplitude combinations is known as a constellation. There is a great deal of similarity between QUAM and the colour subcarrier used in analog television in which the two colour difference signals are encoded into one amplitude and phase modulated waveform. On reception, the waveform is sampled twice per cycle in phase with the two original carriers and the result is a pair of 8-level signals.

The data are randomized by addition to a prs before being fed to the modulator. A serial digital waveform basically contains a train of rectangular pulses. The transform of a rectangle is the function $\sin x/x$ and so the baseband pulse train has a $\sin x/x$ spectrum. When this waveform is used to modulate a carrier the result is a symmetrical $\sin x/x$ spectrum centred on the carrier frequency. Nulls in the spectrum appear spaced at multiples of the bit rate away from the carrier. Further carriers can be placed at such spacings such that each is centred at the nulls of the others. The distance between the carriers is equal to 90 degrees or one quadrant of $\sin x$. Owing to the quadrant spacing, these carriers are mutually orthogonal. A large number of such carriers will interleave to produce an overall spectrum which is almost rectangular as shown in Figure 5.39(a). This approach is known as coded orthogonal frequency division multiplexing or COFDM [5.9]. In typical COFDM systems, several thousand carriers are used in one channel. The bit rate in each is then quite low.

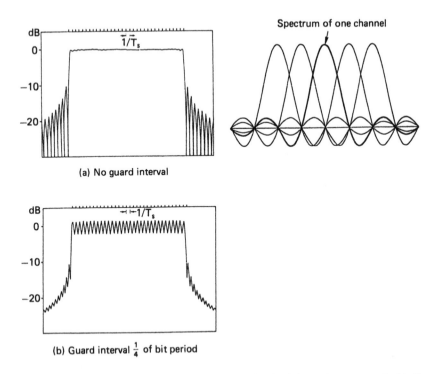

(a) No guard interval

(b) Guard interval $\frac{1}{4}$ of bit period

Figure 5.39 (a) When mutually orthogonal carriers are stacked in a band, the resultant spectrum is virtually flat. (b) When guard intervals are used, the spectrum contains a peak at each channel centre.

In practice, perfect spectral interleaving does not give sufficient immunity from multipath reception. In the time domain, a typical reflective environment turns a transmitted pulse into a pulse train extending over several microseconds. If the bit rate is too high, the reflections from a given bit coincide with later bits, destroying the orthogonality between carriers. Reflections are opposed by the use of guard intervals in which the carrier returns to an unmodulated state for a period which is greater than the period of the reflections. Then the reflections from one transmitted bit decay during the guard interval before the next bit is transmitted. The principle is not dissimilar to the technique of spacing transitions in a recording further apart than the expected jitter. As expected, the use of guard intervals reduces the bit rate of the carrier because for some of the time it is radiating carrier not data. A typical reduction is to around 80% of the capacity without guard intervals. This capacity reduction does, however, improve the error statistics dramatically, such that much less redundancy is required in the error correction system. Thus the effective transmission rate is improved. The use of guard intervals also moves more energy from the sidebands back to the carrier. The frequency spectrum of a set of carriers is no longer perfectly flat but contains a small peak at the centre of each carrier as shown in Figure 5.39(b).

5.17 The data to be transmitted

In digital broadcasting, effectively the channel between the transmitter and the viewer is a multiplexed data link in which some of the data represent picture information and some represent audio. In fact one 'channel' may contain several simultaneous television programmes. Conservation of bandwidth requires that compression is used in digital broadcasting.

The digital output of a single channel sound or picture compressor is called an elementary stream. In many applications, DVB included, several elementary streams will be time-multiplexed into what is known as a transport stream by buffering each elementary stream and squirting the data out at a higher bit rate during the multiplex time slot. An equivalent time expansion process is needed at the demultiplexer.

Numerous housekeeping signals are added to the transport stream, generally called system information. Programme identifiers (PIDs) describe each of the different elementary streams in a transport stream. Data packets are numbered contiguously so that they can be assembled correctly at their destination. The way in which all of these requirements are encoded and handled is known as a protocol.

Each elementary stream must behave like a real-time transmission so the encoder and decoder must be synchronized. Their clocks are locked by a periodic transmission of the programme clock reference (PCR) which the decoder can turn into a stable clock using a numerically locked loop. The exact time at which the pictures or audio samples should be presented to the viewer is conveyed by transmitting periodic time stamps which in conjunction with PCR allow the decoder to rebuild the time axis of the original video and audio.

Index

Milton Keynes UK
Ingram Content Group UK Ltd.
UKHW040057071024
449327UK00019B/629